| RESTRICTED | OP 1673A |

GERMAN UNDERWATER ORDNANCE
MINES

| A BUREAU OF ORDNANCE PUBLICATION | 14 JUNE 1946 |

> This publication is RESTRICTED and shall be safeguarded in accordance
> with the security provisions of U. S. Navy Regulations, 1920, Article 76

MILITARY ARMS RESEARCH SERVICE

DEPARTMENT OF MILITARY BALLISTICS
POST OFFICE BOX 26772
SAN JOSE, CALIFORNIA 95159 U.S.A.

RESTRICTED   OP 1673A

GERMAN UNDERWATER ORDNANCE
MINES

14 JUNE 1946

This publication is RESTRICTED and shall be safeguarded in accordance
with the security provisions of U. S. Navy Regulations, 1920, Article 76

NAVY DEPARTMENT

BUREAU OF ORDNANCE

WASHINGTON 25, D. C.

RESTRICTED   14 June 1946

ORDNANCE PAMPHLET 1673A

GERMAN UNDERWATER ORDNANCE: MINES

    1.   Ordnance Pamphlet 1673A provides basic information on each type of German sea mine used or in development during World War II. It is not an exhaustive analysis of these mines, but an overall survey of the German discoveries, experiences and achievements.

    2.   The only attempts to compile existing information on German mines into published form were in OP 1330 and OP 898, both of which were limited in scope. OP 1330 was intended solely for mine disposal officers and OP 898 was an identification manual for general service personnel. This publication includes some of the information from each of these previous pamphlets. Other information has been taken from letter and technical reports prepared by the Naval Technical Mission in Europe, intelligence reports forwarded by the Commander, Naval Forces in Europe and Commander, Naval Forces North African Waters, and field intelligence reports from mine disposal officers assigned to Mobile Explosive Investigation Units Numbers 2 and 3. These reports are listed in the Bibliography.

    3.   The information relating to several of the mine items described herein is incomplete for one or more of the following reasons:

        a. No specimens were found.
        b. No documents relating to it were available.
        c. Reliable information could not be obtained through interrogation.
        d. Specimens have not as yet been analyzed by the cognizant technical activities.

    4.   This publication is RESTRICTED and shall be safeguarded in accordance with the security provisions of U. S. Navy Regulations, 1920, Article 76.

                                              G. F. HUSSEY, JR.
                                              Vice Admiral, U. S. Navy
                                              Chief of the Bureau of Ordnance

TABLE OF CONTENTS

Chapter 1 - History

	Page
Magnetic Units	1
Degaussing	1
Acoustic Units	1
Research after 1941	2
Pressure Units	2
Experimental Units	2
Optical	2
Cosmic Ray	2
Infra Red	2
Underwater Electrical Potential	2
Gravitation	2
Wellensonde (Wave Probe)	2
Combined Units	2
Auxiliary Devices	3
Arming Clocks	3
Period Delay Mechanisms	3
Sterilizers	3
Pausenuhr	3
Twelve-Hour EW	3
Prevent-Stripping Mechanisms	3
Raumschutz	3
Naval Mine Cases	3
Contact Mines	4
Contact-Influence Mines	4
Anti-Asdic Research	4
Conclusion	4

Chapter 2 - German Mine Organization

Functions of Mine and Minesweeping Group (SVK)	6
Personnel of Mine and Minesweeping Group (SVK)	6
Facilities of Mine and Minesweeping Group (SVK)	6
Trials Group (SEK: Sperrwaffenerprobungskommando)	6
Functions of Test Station (E-Stelle)	7
Personnel of Test Station (E-Stelle)	7
Facilities of Test Station (E-Stelle)	7
Coordination between the Naval Mine and Minesweeping Group (SVK) and Luftwaffe Test Station (E-Stelle)	7
The Effect of Allied Bombing on the Mine Program	8
Organization Chart - SVK	8
Organization Chart - Luftwaffe	9

Chapter 3 - Critique

Critique	10
Technical Development	11

Chapter 4 - Contact and Moored Influence Mines

The German EM (Einheitsmine) Mine Series	14
The EMA Mine	14
The EMB Mine	15
The EMC I Mine	15
EMC I - EMC II - EMC II (Upper Antenna)	15

	Page
EMC II Mine	16
The EMD Mine	18
The EME Mine	20
The EMF Mine	20
The EMG Mine	22
The EMH Mine	22
The EMI Mine	23
The EMK and EMU Mines	23
The EMR and EMR/M Mines	23
The EMS Mine	25
The FM Mines	25
FMA Mine	25
FMB Mine	25
FMC Mine	25
The OMA Mines	26
The UM Mines	28
The UMA Mine	28
The UMB Mine	30
Base Plates	32
Table of Base Plates	37
Table of Contact Mines	42

Chapter 5 - Aircraft Mines - SVK

The LM Mines	46
The LMA Mine	46
The LMB Mine	48
The LMC Mine	51
The LMD Mine	51
LM - ZUS Z (34) B Bomb Fuze for LMB	51
Parachutes for the LMA and LMB Mines	55
LMF Mine	56

Chapter 6 - Aircraft Mines - Luftwaffe

The FMC Mine	58
The BM Mine Series	59
BM 1000 I	60
BM 1000 II	63
BM 1000 C	63
BM 1000 H	63
BM 1000 J - J I - J II - J III	63
BM 1000 M	64
BM 1000 T	64
BM 500	65
BM 250	65
Bugspiegles (BS 1 and BS 2)	65
Bugverkleidung (BV 2 and BV 3)	65
Schutzhaubes	66
SH 1 and 2	66
SH 8 and 9	66
SH 7	66
SH 11	66
Leitwerke (LW)	66
Assembly Plan for BM 1000 Mines	70
BM 1000 Parachutes	71
LS 1	71
LS 3	73
LFS-08	73
Special Accessories used with BM 1000	75
Rheinmetall Bomb Fuze - Type 157/3	75
Master Switch	75
Fuse Delay Switch	76

RESTRICTED

iii

GERMAN UNDERWATER ORDNANCE—MINES

OP 1673A

	Page
Chapter 7 - Submarine-Laid Mines	
The SM Mines	77
The SMA Mine	77
The SMB Mine	78
The SMC Mine	78
The TM Mines	78
The TMA I Mine	78
The TMB Mine	80
The TMC Mine	81
Chapter 8 - Controlled Mines	
The KMB Mine	83
The RM Mines	83
The RMA Mine	83
The RMB Mine	84
The RMC Mine	85
RM Mines used as Influence Mines	85
RM Mines used as Controlled Mines	85
The RMD Mine	85
The RME Mine	86
The RMH Mine	86
The RMH used as an Influence Mine	87
The RMH used as a Controlled Mine	87
Table of Influence Mines	88
Chapter 9 - Special Purpose Mines	
The HM Mines	94
General	94
The HMA Mine	94
The HMB Mine	94
The HMC Mine	96
The KM Mines	96
The KMA Mine	96
The MTA Mine	97
The BM 1000 F Mine	97
The BM 1000 L Mine	98
Winterballoon	98
Wasserballoon	99
Chapter 10 - Sweep Obstructors	
Introduction	100
Explosive Conical Floats (Sprengboje C and Sprengboje D)	100
Static Conical Sweep Obstructor (Reisboje)	103
Aircraft-Laid Sweep Obstructors (BRA - BRB)	103
BRA	103
BRB	104
Tables of Anchors	105
Chapter 11 - Influence Mine Units - SVK and Luftwaffe	
Section 1 - Influence Firing Devices - Tables	
Influence Mine Units - Navy	109
Influence Mine Units - Luftwaffe	114
Operational Ground Influence Mines - Navy	117
Operational Moored Influence Mines - Navy	118

	Page
Section 2 - Magnetic Units	
M 1 Unit and Modifications	119
M 1 Unit	119
M 1s Unit	119
M 1r, MA 1r, and MA 1ar	121
M 1 Unit Circuit - Operation	121
M 1 (1st Revision) Unit Circuit - Operation	127
M 1 (2nd Revision) Unit Circuit - Operation	127
M 2 Unit	127
M 2 Unit Circuit - Operation	127
M 3 Unit	129
M 3 Unit Circuit - Operation	129
M 3 (Revision) Unit Circuit - Operation	130
M 4 Unit	130
Unit Construction	130
Latitude Adjustment and Arming Process	130
M 4 Used in Combination with Another Unit	133
M 5 Unit	133
M 101 Unit	138
M 101 Unit Circuit - Operation	138
M 101 with Three-Place P.D.M.	141
M 101 with Ten-Place P.D.M.	143
M 103 Unit	143
M 103 Operation	143
Section 3 - Acoustic Mine Units	
A 1 Unit	144
A 1 Unit - Operation	144
A 1 (Allied - A Mk II) Unit	148
A 1 (Allied - A Mk III) Unit	150
A 1st - A 2st (Allied - A Mk VI) Circuit Operation	151
Microphones	151
Microphone Circuit	152
Basic Systems	154
System 1	154
System 2	155
System 3	155
System 1 - A 1, A 1st, A 2, A 2st, A 3	155
A 104	156
A 104 Circuit - Operation	159
System 2 - A 4, A 4st, A 105, A 105st	161
A 105 (Circuit - Operation)	163
System 3 - A 7 and A 107	163
Relay Reactions	165
A 107tc	169
Circuit Components	169
Section 4 - Magnetic-Acoustic Combination Units	
MA 1, MA 1a, MA 1st, MA 2, and MA 3 Units	169
MA 1	169
MA 1a	171
MA 1st	177
MA 2	177
MA 2 with ZK IIc	178
MA 3	178
MA 101 Unit	179
MA 105 Unit	182

Section 5 - Subsonic Units

	Page
AT 1, AT 2, and AT 3 Units	182
Magnetophone	182
Magnetophone Construction	182
AT 1	185
Acoustic Triggering Circuit	185
AT 2 Amplifier (Subsonic)	186
AT 3 Amplifier (Subsonic)	186
"Hell" Doppelschwinger	186

Section 6 - Supersonic Units

	Page
AA 4 and AA 106 Units	183
AA 4 Unit	190
AA 4 Circuits	190
AA 106 Unit	191
Acoustic Triggering Circuit	191
Supersonic Receivers	191
Float and Float Release	192
P.S.E. and Anti-Leak Device	192
AE 1 and AE 101 Units	192
Possible Countermeasures to the Mine	194
Twin Receiver Mine A 106	195
The Magnetostriction Receivers	195
The Float	195
The Circuit	196
Preliminary Switching On	196
The Triggering Circuit	196
Amplifier A for Horizontal Receiver	197
Amplifier B for Vertical Receiver	197
Timing Circuit	197
Anti-Freezing Switch	198
Sea Trials	198
Echo-Sounder Mines Types AE 1 and AE 101	198
The Magnetostriction Transducers	198
The Float	201
The Mineshells	201
The Circuits	201
Arming of the AE 1	201
Arming of the AE 101	201
Acoustic Trigger Circuit	201
The Rotary Converter	201
The High-Speed Camshaft	202
The Low-Speed Camshaft	202
The Lowest-Speed Camshaft	202
Transmitter and Firing Circuits	202
The Receiver Circuit	203
The Amplifier	203
The Trigger Circuit	203
AVC Amplifier and anti-sweep Circuit	205

Section 7 - Pressure and Pressure-Combination Units

	Page
Pressure Mine Units and Detecting Devices	205
Pressure-Detecting Device	205
General	205
Description	205
Operation	206
Pressure Components D 2 and D 102	206
Operation of D 103 Unit	208
Characteristics of the D 113	209
Characteristics of the D 123	209
Characteristics of the D 133	210
AD 104 Unit	210
Operation	210
DA 102 Mine Unit Series	212
The Microphone	212
DA 102 Operation	212
DA 112	213
DA 132	213
DA 162	213
DA 244	213
DA 122, DA 142, and DA 152	213
DM 1 Unit	214

Section 8 - Experimental Units

	Page
Seismik Mine Unit	216
AJ 102 and AJD 102 Units	216
JDA 105 Unit	220
AMT Units	220
Mounting	221
ZR II Mechanism	221
Blocking	221
Operational Characteristics of AMT 1	221
Operational Characteristics of AMT 2	222
S 101 Unit	222
MDA 106	224
Cosmic Ray Unit	224
Wellensonde Unit	226
Gravitation Mine Unit	227
Elecktroden Effekt	228
Development of Optical Mine-Firing Mechanisms	229
Luftwaffe Units	229
SVK Test Model 1	229
SVK Test Model 2	230
SVK Test Model 3	230
SVK Test Model 4	230
SVK Test Model 5	230
SVK Test Model 6	230
Conclusions	230
Luftwaffe Forelle Unit	231
Induction Mine Units	231
BMA II	232
BMA III	232
J I	232
J V	233
Characteristics of Influence Mines - Navy	234
Sweeping Characteristics of Influence Mine Units - Navy	236

Chapter 12 - Clocks and Associated Devices

	Page
Introduction	237
Arming Clocks	237
UES I	237
UES II	237
UES IIa	237
Clock Starter Plate	237
UES II and UES IIa with LiS	237
UES II and UES IIa with Vorkontakt	238
Sterilizers (Zeit Einrichtungen)	241
ZE	241
ZE I	241
ZE II	242
ZE III	242

	Page
ZE III for 360 Days	243
ZE IV	243
ZE IVa	244
ZE V	244
ZE VI	244
Period Delay Mechanisms (Zahl Kontakt)	245
ZK I	245
ZK II	245
ZK IIa	245
ZK IIb	245
ZK IIc	245
ZK IId	245
ZK IIe	245
ZK IIf	247
Miscellaneous Clockwork Mechanisms-	247
Pausenuhr (Arming and Disarming Clock)	247
Entscharfer Werke (EW) (Disarming Clock)	247
Verzogerungs Kontakt (VK) (Delay Contact)	247
Verzogerung Werke (VW) (Delay Clock)	248
Period Delay Mechanisms	250
Mechanical P.D.M.	250
Electrical P.D.M., - Fuse-Delay Switch	250
Automatic Latitude-Adjustment Devices	250
Mechanical A.L.A.	250
Magnetic A.L.A.	252
Electromagnetic A.L.A.	252
Prevent-Stripping Equipment (Geheimhaltereinrichtung)	255
LMB P.S.E.	255
TMA P.S.E.	255
BM 1000 P.S.E.	257
Mechanical P.S.E.	257
Sea-Cell P.S.E.	257
ZUS-40 Anti-Withdrawal Device	257

	Page
Chapter 13 - Miscellaneous mines	
KK Mine	259
RK Mine	259
K TR Mi Mines of the Wehrmacht	261
K TR Mi 38	261
K TR Mi 41	264
Bibliography	
Technical Reports - Naval Technical Mission in Europe	265
Mine and Mine-Unit Documents forwarded from Germany	268
Miscellaneous Reports	270
Addenda	
Personalities	274
SVK	274
E-Stelle (Luftwaffe)	275
Others	275
Firms Working on Mine Development	276
Research Institutes Working on Mines	277
Abbreviations	
Abbreviations, German - Translations	278
Mine Unit Designations - Allied - German	280
Mine Designations - Allied - German	280

Chapter 1

HISTORY

Beginning in the early 1920's, the German Navy developed an extensive marine mine research program, and by the outbreak of World War II possessed a number of revolutionary types ready for mass production. These included seven contact mines, eight influence mine cases, and three magnetic units. Among them were ground and moored mines suitable for laying by aircraft, surface vessels, and submarines.

The German Navy had already perfected highly efficient contact mines during the first World War, but Allied countermeasures seriously limited their effectiveness. Consequently, in 1923, research began upon mines operating on non-contact principles, and was first directed toward the development of a magnetic unit.

MAGNETIC UNITS

Magnetic units of either the induction or the needle type were feasible; but, since Germany was not self-sufficient in the copper and nickel essential to induction units, work centered upon the needle unit. The first needle type, E-Bik, was completed in 1925, and in subsequent years it was improved and adapted to ground and moored mines. By 1939 the M 1 (unipolar), M 2 (unipolar), and M 3 (bipolar) magnetic units were ready for operational use.

During the war, the Navy continued research for improvements to the M 2, and M 3, and for methods to keep ahead of Allied countermeasures. This resulted, before the end of the war, in the M 4 (unipolar) and M 5 (bipolar). The M 4 was an improvement of the M 3 type, designed for use in either moored or ground mines in combination with other units. It possessed a maximum sensitivity of 2.5 mg., and contained a device which reset the unit automatically when it was actuated or disturbed. The M 5, a small, improved M 1 designed for use in ground mines in combination with other units, was abandoned in favor of the more satisfactory M 4. The latter was used operationally; the M 5 was not.

Raw material shortages prevented large-scale German production of induction-type units. A limited research program failed to develop substitutes for copper windings and nickel rods. Aluminum windings on high permeability steel rods were tried; but, despite the use of up to eight rods in a single mine case, the loss in sensitivity was too great.

Attempts were also made to develop amplified induction units for use in combination with other units. The Luftwaffe, which was in a favored position throughout the war, obtained small amounts of copper and nickel to develop the AJD 102 and JDA 105 combination induction units. Neither one of these was used operationally.

DEGAUSSING

Concomitant with the development of magnetic mine units, the German Navy pressed an intensive countermeasure program designed to defend itself against the magnetic mines which it thought the British had developed. All German warships were degaussed by 1939, and the Merchant Marine by 1940.

ACOUSTIC UNITS

The Germans assumed erroneously that the British, who had laid the first magnetic mine, would have developed sweeping techniques and established degaussing procedures to counter it. Consequently, in 1938, when the magnetic mine program was well advanced, they commenced research upon acoustic mine firing mechanisms.

Because of the unexpected initial success of magnetic mines, acoustic research continued at a low pace until the British inaugurated effective countermeasures in the spring of 1940. The German Navy in May assigned a sonic acoustic project with highest priority to Dr. Hell Firma. By September Dr. Firma had developed the A 1 unit ready for operational use. Thereafter, he made steady improvements in it, and sought other types of acoustical units.

The A 3 acoustic unit for use with the EMF and SMA moored influence mines proved unsatisfactory. In one test, 100 of these units, fitted in EMF cases, were laid in the Kattegat, and almost all of them simultaneously prematured. Dr. Firma sought to remedy the defects of the A 3 in the A 7, which was in its final test stages when World War II ended.

Other research developed the AT subsonic units (also known as AA units) and the AE supersonic units. The AT units were used operationally from 1942 to the close of the war, but the AE units did not progress beyond the advanced development stage. The AE, which functions satisfactorily in greater depths than other acoustic types, was intended for use in moored mines to be laid in the relatively deep water off the American coast.

Other acoustic units embodying devices designed the A 4, AA 4 and Seismik. The first two differed from the A 1 principally in that the A 4 depended on the rate of the change of the sound level and the AA 4 depended on the directional characteristics of the sound. The Seismik (Sismograph Microphone) consisted of a simple electrical circuit utilizing a carbon-button microphone mounted within a modified D 1 pressure unit and intended to respond at the subsonic levels of 5-8 cps. The circuit was intended for use in combination with other mechanisms such as the M 4, A 4, and D 2.

RESEARCH AFTER 1941

By 1941, British countermeasures against magnetic and acoustic mines were so effective that the Germans began a new research program to develop units using new firing principles, combined units, and auxiliary devices calculated to hinder or defeat British sweeping methods.

PRESSURE UNITS

The first objective was realized by 1943, when the pressure firing mechanism D 1 and D 101 were completed and readied for operational use. These were used with the M 1 and A 104 units to form the combined units DM 1 and AD 104. Subsequently, the D 2 and D 102, designed to provide against actuation by nearby explosions, were developed for use in the same combination as their prototypes.

A series of uncombined pressure units, the D 103, D 113, D 123, and D 133, were designed for rivers or other relatively smooth waters. Only the D 103 was used operationally; the others were still under development at the end of the war.

A general rule prohibiting the employment of mines for which no countermeasures existed prevented pressure units from being employed until after the Allied Invasion of France in 1944. The military situation by then was so grave that the Germans even laid early defective types with natural rubber bags, which allowed the air to escape in as little as three weeks after the unit was underwater. This flaw, discovered only a short time before the invasion, was remedied by substitution of leakproof synthetic rubber bags.

EXPERIMENTAL UNITS

An over-all plan to examine all physical laws that might be applied to influence mine units led the Germans to undertake varying degrees of experimentation with the following types of units: optical, cosmic ray, infra red, UEP (Elektroden Effekt), gravitation, and Wellensonde (wave probe).

Optical. The Navy devised a number of experimental optical units. Those designed for use in the open sea presented serious difficulties, and none was perfected before the end of the war. However, a river unit used operationally in the latter stages of the war against river bridges, achieved some success. All these units utilized photoelectric cells so arranged that decreases in light of prescribed intensity and rate would actuate the unit.

Cosmic Ray. Preliminary development was begun upon a cosmic ray unit that would operate on the increase of the underwater cosmic ray level caused by the passage of a ship. The earliest experiments were conducted with twenty-four standard Geiger-Müller counter tubes cast into the explosive of an LMB mine case. Difficulties in obtaining a satisfactorily high potential supply and in regulating this potential slowed the project to such an extent that it was still in little more than the "idea stage" at the war's termination.

Infra Red. Although the Germans made inquiry into the possible application of the underwater action of infra red rays, no information on the progress in this field was obtained by U.S. Naval investigators, and, to date, no reports have been received from British sources.

Underwater Electrical Potential (Elektroden Effekt). During the course of World War II the German Navy discovered that the passage of a ship created an electrical current in the water which could be detected by copper electrodes placed on the sea bottom. They termed this phenomenon the "Electroden Effekt" and sought to develop a mine unit that would operate on this principle. Progress was slow, and at the close of the war the investigation was still in the preliminary stages.

Gravitation. The Askania Werke, Berlin, attempted to create a mine mechanism that would work on a principle similar to that of the Askania Gravity Balance. The Werke made certain calculations in conjunction with the Geophysical Institute of Potsdam, but the idea made relatively insignificant progress.

Wellensonde (Wave Probe). The Germans sought to utilize the distortion which passing vessels would effect upon high frequency alternating currents emanating from a mine case, in order to fire a unit. The principle was similar to the U.S. Navy's Electrical Discontinuity Discriminator, but the German application never progressed beyond the experimental stage.

COMBINED UNITS

To thwart British countermeasures and give greater life and sensitivity to units, the German Navy developed a series consisting of a combination of two or more firing mechanisms, each of which operated on a different principle. The earliest of these, the MA 1, built in 1941, combined the M 1 magnetic and A 1 acoustic units. Improved versions known as the MA 1a, MA 2 and MA 3 followed. The Luftwaffe made a similar combination, the MA 101, with improvements designated MA 102 and MA 105.

In 1942 the pressure-magnetic DM 1 went into production; the following year, the pressure-acoustical AD 104. These early pressure combinations were followed by the DA 102 pressure-acoustical series of seven different units. Of these, only the DA 102 was used operationally.

Another important series of combined units, the double acoustics, utilized a sonic acoustic system for triggering the subsonic or supersonic systems. These units were known as the AT 1, AT 2, AT 3 and AA 106 (All subsonic) and the AE 1 and AE 101 (both supersonic).

By May 1945, three-unit combinations were reaching the operational stage. They consisted of the MDA 106 (Magnetic-pressure-acoustic), JDA 105 (induction-pressure-acoustic) and the AMT 1 and 2 (acoustic-magnetic-subsonic). Two miscellaneous combined units which never reached an operational stage were the DS 1 (pressure-seismik) and the JD 102 (induction-pressure).

AUXILIARY DEVICES

Arming Clocks. As a necessary accessory to the first influence mine units, the German Navy designed a six-day arming clock, the UES I. The primary purpose of the arming clock was to allow influence mines of the ground type to settle securely in the bottom prior to arming. Secondarily, it was used to hinder sweeping operations. The range of the clock was from one-half hour to six days, and the settings were always for the maximum period consistent with the military objective. Various improvements were made both prior to and during the course of the war, but basic operation was never altered.

Late in the war the German Navy introduced a new type of arming clock, the ZE III. This clock could be set from five to two hundred days, and could be utilized either for arming or for disarming. A similar 360-day clock, also known as the ZE III, was under development at the close of the war. The only other delaying arming clock used was an eighty-hour type employed with the BM 1000 mines to permit proper orientation of the case prior to arming.

Period Delay Mechanisms. Another important series of clockwork mechanisms used in German mines were the Period Delay Mechanisms ZK I, ZK II and ZK IIa through ZK IIf. The mechanisms were so designed that from one to eighty-five actuations within prescribed time periods were necessary to fire the mine. The first of the mechanisms, which possessed a span of only six actuations, was intended to defeat the practice of having minesweepers safeguard outgoing vessels by preceding their passage from port. The last model, the ZK IIf, which could be set up to eighty-five, was designed to make clearance sweeping extremely burdensome.

Sterilizers. The third large group of clockwork mechanisms consisted of seven different types of sterilizers, with maximum time periods as follows: ZE (80 days), ZE I (80 days), ZE II (6 days), ZE III (200 days), ZE III (360 days), ZE IV (45 days) and ZE IVa (60 days). In addition, a 200-day electrolytic cadmium cell sterilizer was developed.

All the sterilizers were used in various German mines to limit the life of minefields in accordance with tactical requirements. They were widely used to permit the replenishment of E-boat laid minefields off the English coast.

Pausenuhr. Several clockwork mechanisms served special purposes. The most important was an 18-day clock, the Pausenuhr, which armed and disarmed a mine once in every 24 hours. The German Navy developed this clock after it observed that the British normally made morning sweeps after minelaying sorties and allowed traffic to resume by midday.

Twelve-Hour EW. Another clock, the twelve-hour EW, was used with the M 3 unit in moored influence mines. The EW tested the mine circuit for a period of up to twelve hours after laying, and scuttled the mine at the end of the set period if it planted improperly or was otherwise disturbed.

Prevent-Stripping Mechanisms. To protect influence mines from capture, the Germans devised a variety of mechanisms commonly referred to as "booby-traps" or "Prevent-Stripping Mechanisms." They consisted of specially designed bomb fuzes to explode aircraft mines that fell on shore, photoelectric cells which fired the mine if it were exposed to light by the removal of the unit dome, sea cells to explode the mine if the unit was exposed to moisture, a variety of anti-withdrawal and anti-removal devices, and a unit to fire the mine if it were moved into shallow water or inadvertently laid in tidal flats. These devices were used widely in the early phases of the war. However, after several accidents involving the loss of German mining personnel, they fell into disfavor. During the latter part of the war they were used infrequently.

Raumschutz. When the Germans acquired complete information on Allied magnetic sweeping procedure through the capture of a BYMS off Leros, they undertook development of Raumschutz (area protection) to defeat the LL-type sweep. These units, for use with the M 1, MA 1 and MA 1a mines, were to be designated M 1r, MA 14 and MA 1ar. They were in production in 1945.

Raumschutz was a rubber-covered cable 165 feet in length with one copper electrode secured to the end and another mounted on, but insulated from, the mine case. In operation, the sea current produced by an LL-type sweep was picked up by the electrodes and, through a sensitive relay, the mine was rendered passive for the duration of the sea current plus a predetermined period. According to reports, the Germans attempted to fit Raumschutz to aircraft-laid mines using the MA 105 unit. However, considering the nature of the device, probably the inherent mechanical difficulties were insurmountable.

NAVAL MINE CASES

By May 1945, the German Navy and the Luftwaffe had either laid or undertaken the development of an imposing array of 96 different types of naval mine cases. This total does not take into consideration captured foreign mines which the Germans used. The mines fall into two separate groups -- contact and influence.

Contact Mines. The Germans started World War II with seven different types of moored contact mines: the EMA, EMB, EMC, EMD, FMB, FMC, and UMA. The EMA and EMB, identical except for the weight of charge, were World War I mines designed for laying by submarines. A limited quantity were laid during World War II out of stocks remaining on hand. The Japanese JA mine, used operationally in the Pacific after 1941, was a copy of the EMA.

The German Navy developed the remaining five types between 1923 and 1939. It placed especial emphasis upon the EMC, which was the most widely used and the most adaptable. During the course of the war, the Navy made major changes in chain moorings; added cork-floated snag lines, mounted antenna and mechanical cutters on the mooring cable; and in other ways improved these types to remedy defects or conform with changing military requirements.

In addition, a number of new contact mines were developed. The UMB, a larger UMA, was designed; an aircraft-laid, moored, contact mine, the BMC, was introduced; the EMS series of drifting, decoy, contact mines were readied for operational use; the OMA series of moored, surface-contact mines and the EMG shallow-water, constant-depth assembly appeared. This group of moored contact mines was fortified by the development of a ground contact mine made of concrete and known as KMA. With this mine arsenal, the Germans had a series of diversified Naval mines adequate for composing a contact minefield that would meet the requirements of any given tactical need.

Contact-Influence Mines. Two interesting contact-mine developments were undertaken during the war. The first of these was the design of two combination contact-influence mine cases, the EMK and EMU. These mines were intended to overcome the following shortcomings found in previous German moored mines:

1. In deep water, hydrostatic pressure sometimes prevented arming by counterbalancing the pull of the mooring cable.

2. In shallow water, rough seas caused excessive arming and disarming.

3. The use of explosive containers within the mine reduced the damage radius and served to render proper mine orientation more difficult.

4. The plummet-type standard surface anchor was not suitable for delayed-rising mines.

Since this development was assigned a low priority it proceeded at a slow pace and was never completed.

ANTI-ASDIC RESEARCH

The Germans began another development when the relatively small mine damage in the initial amphibious assaults at Anzio and Salerno indicated that the Allies had perfected a method of detecting moored mines by ASDIC. The U-Boat Command for some time had been seeking an answer to the problem through special paints and coats of rubber. The Mine Command tested and rejected these. Very late in the war it hoped to eliminate rather than merely reduce the response obtained from cases with a metal core through design of an all sponge-rubber case with a minimum of rubber fittings.

CONCLUSION

When the Germans launched their research for a magnetic mine unit, they simultaneously undertook the development of a mine case to house it. The earliest of those cases were the RMA and RMB ground mines, both of which were hemispherical in shape, of all-aluminum construction, and designed for laying by surface vessels. The hemispherical design was to insure proper orientation of the magnetic unit after planting. When experimentation showed that the case tended to sled on laying, a specially designed float was added to the mine. The purpose of the float was similar to that of a drogue, i.e. to slow the descent of the mine in water and to prevent sledding and tumbling.

During the war, additional mines of the RM series were developed. This series consisted of surface-laid ground mines which could be utilized as influence and/or controlled mines. These were the RMD, RME, and RMH. The RMD could be fitted with any of the various firing units; the RME was for use in rivers with an M 1 unit; and the RMH was a wooden-box sea mine of simple design which also housed an M 1 unit. This RMH was intended for local fabrication, so that overtaxed transportation facilities could be partially relieved from carrying bulky and heavy mine cases over long distances.

After completing the RMA and RMB, the Germans sought to exploit the potential value of influence mines that could be laid by aircraft. Accordingly, they developed the parachute mines LMA and LMB. These were ground mines, cylindrical in shape and of all-aluminum construction. They were used very widely during the war, with satisfactory results except for one important factor; the maximum laying speeds and altitudes were too low. This led the Luftwaffe to push the perfection of a high-altitude, high-speed mine that would provide greater safety for the laying aircraft. The answer found was the Bomb-Mine 1000, which is discussed in later paragraphs.

The LMA and LMB received several interesting wartime modifications. When British airpower and antiaircraft fire made aircraft mining extremely hazardous, the mines were changed so that they could be laid by E-Boats, and redesignated LMA/S and LMB/S. They differed from the air types only in the type of tail used and the elimination of the bomb fuze. A further innovation in all these mines was the substitution of pressed paper (Prestoff) for aluminum in the

HISTORY

fabrication of the mine case, in order to reduce the high cost and to forestall any future aluminum shortage. It was difficult to keep these types, the LMA/F and LMB/F, watertight, but, as they were found stored at operational airfields in France and Belgium, they were in use or ready for use.

The LMA and LMB mines presented an additional problem. Since they were of the ground influence type they could not be employed in depths of up to 1,000 feet. Later a modified LMF, the LMF/S was introduced for laying by E-Boats.

The Luftwaffe, being unsatisfied with the LM mines because they necessitated low-altitude drops at low speeds, pushed the development of their cylindrical, manganese steel bomb-mines, the BM 1000 series. By 1945 they had developed mines that they could drop from heights of up to 21,000 feet at speeds of over 400 m.p.h. This was achieved by the use of break-away flat noses, small parachutes, and other accessories. Of thirteen different types of bomb-mines, five were used operationally. One, the BM 1000t, was a moored mine intended to attain the same results as the LMF. However, this mine proved unsatisfactory during dropping tests and was abandoned in 1944.

To round out their influence mine program, the Germans developed a variety of ground and moored types for laying from submarines. The earliest were those which could be laid from submarine torpedo tubes, the TMA (moored) and the TMB (ground). In 1939 the TMC, a larger version of the TMB, was developed to meet field requests for a submarine-laid mine with a heavy charge. TMC housed over 2,000 lb. of hexanite.

When these mines had been completed, the development of a mine torpedo (MTA) was undertaken. This project was successfully completed by 1942, and mines of this type were used operationally during the war. The mine was intended for use in harbors, to be laid by being fired from submarines standing off shore. The advantages were that it reduced the risk of detection by harbor defenses and permitted safe replenishment of fields already laid.

The other series of submarine-laid mines were all of the moored type for laying from the vertical shafts of special minelaying submarines. The first of this series was the SMA, which was laid off the Americas in fields off Halifax and Panama in 1942-1943. In 1943 a clock was added to the SMA anchor to obtain a delay of up to 60 days in separation of mines and anchor. This modified assembly was called the SMC. The development of an additional type of SM mine, the SMB, was undertaken prior to the war but never progressed beyond the preliminary design stage because of the low priority assigned to the project. This mine was intended for use in depths of up to 9,000 feet, especially off the American coasts.

Simultaneously with the development of the TM mine series, the German Navy perfected a moored influence case for surface laying, the EMF. This assembly was designed for use in depths of up to approximately 1,600 feet as opposed to the maximum of about 125 feet for the ground influence mines; it was used extensively during the war.

Chapter 2
THE GERMAN MINE ORGANIZATION

The earliest German establishment charged with the technical development and the testing of Sea Mines was the Navy's Mine and Mine-Sweeping Group (SVK: Sperrversuchskommando) which fitted into the over-all German Mining Program as shown in Figure 1. In 1938, the Luftwaffe had decided that the LM-type parachute mines developed by SVK would not meet future tactical requirements and contracted with the private firm of A.E.G. for the development of a bomb-mine. However, SVK remained in complete control of mine development until 1941, when the Luftwaffe created a separate and independent organization known as the Test Station (E-Stelle: Eprobungstelle) for the purpose of pushing the development of the bomb-mine. (The E-Stelle was established within the Luftwaffe, as shown in Figure 2.). In 1943, SVK was shorn of further responsibilities with the formation of an independent Navy testing establishment known as the Trials Group (SEK: Sperrwaffenerprobungskommando).

FUNCTIONS OF MINE AND MINE-SWEEPING GROUP(SVK)

Until 1941, SVK was charged with the development of all German Naval mines, including aircraft types. The specifications and requirements for the mines were laid down by the General Staff, after consultation with SVK. If, during the progress of the work, it was found that original requirements could not be met, the specifications were modified accordingly. In addition, SVK furnished first drafts of instructional and operational pamphlets, as required by Mine Inspection SI (Sperrwaffeninspektion), and made manufacturing drawings of the mines and mine components for contracting firms.

PERSONNEL OF MINE AND MINE-SWEEPING GROUP(SVK)

The Personnel of SVK was under the supervision of Kdr.Kapt. Zur See Bramesfeld, and were divided into two departments (1) Naval and (2) Technical. The Naval Department was assigned approximately 500 Naval personnel who were allocated, as needed, among the Department's four sections; (1) Mines, (2) Influence Sweeps, (3) Depth Charges, etc., (4) Sweeps, and a flotilla consisting of about 200 vessels ranging in size and purpose from steamships to motor boats. The Technical Department consisted of about 450 personnel, about 60 of whom were in the scientific and design division, which was divided as follows: (1) Mines, general mechanical, (2) Sweeps, (3) Influence devices and physical fields, (4) Depth charges and aircraft mine cases.

The drafting rooms and shops were allocated 50 and 300 personnel respectively, the remaining personnel being administrative and clerical. Some Technical personnel held special Naval rank, while others remained in a civilian status. In practice, the scientific, technical, and design work was handled by the Technical Department and the Marine experience added by the Naval Department.

FACILITIES OF MINE AND MINE-SWEEPING GROUP(SVK)

The principal buildings comprising the SVK installation at Kiel were as follows:

1. Main laboratory and administration building, including a large wing for drafting, files, and reproductions.

2. Main shop buildings in the form of a quadrangle with the magnetic laboratory (all-wood construction) in the center of the quadrangle.

3. Mine School building and barracks.

4. Foreign Mine Museum building.

5. Assembly and storage building beside the docks.

6. Tank building similar to that of NOL, Navy Yard, Washington, D.C.

7. Miscellaneous service buildings.

The major testing equipment available to SVK was as follows:

1. Physical testing machines of the type normally found in metallurgical laboratories.

2. Pressure tanks of various sizes, including types large enough to accommodate mine cases.

3. Several laboratory ships, suitable for testing mines on the sea bottom.

4. Electrical and acoustical laboratory instruments.

5. Drop testing equipment and shock testing gun.

Since SVK preferred to make field measurements of mine units on the sea bottom, they made no attempt to perfect large acoustical tanks or simulation equipment for testing magnetic units.

TRIALS GROUP(SEK:SPERRWAFFENERPROBUNGSKOMMANDO)

In the Fall of 1943, the SEK was established under the command of a Fgt. Kpt. Broeckelmann for the purpose of testing mine material for serviceability from the seaman's point of view.

GERMAN MINE ORGANIZATION

This command was based at Kalungborg, Denmark and consisted of a flotilla of mine layers, mine sweepers and work boats. Approximately 700 officers and ratings comprised the staff. Departments were set up for each type of mine and gear, such as controlled mines, moored mines, ground mines, mechanical mine-sweeping gear, etc. An aircraft mine section, headed by an Air Force officer, was created; but, on account of the shortage of planes, fuel and personnel, little aircraft testing was carried out, the task being left to the Luftwaffe.

Apparently, there was no set pattern for acoustic or magnetic trials of mines. Target ships provided the actuation under the supervision of a specialist officer, who had familiarized himself with the unit being tested by working together with SVK. Tests were designed to simulate field conditions as closely as possible.

Trials Group (SEK) reported the trial results to Mine Inspection (SI) and made its own recommendations; if the item was already operational, SEK could advise discontinuance. Drafts of publications prepared by SVK and covering the item under test were checked by SEK from an operational viewpoint.

FUNCTIONS OF TEST STATION (E-STELLE)

The function of E-Stelle, Section E-7 (Mines) was mainly the testing of aircraft mines. However, this group was also responsible for the coordination of development work being carried on by various outside agencies and for some independent development work. (E-Stelle differed from SVK in that practically all of the former's research, development, and drafting work was done by outside agencies which consisted mainly of private manufacturing firms).

PERSONNEL OF TEST STATION (E-STELLE)

The personnel of E-Stelle were under the immediate control of a Luftwaffe Captain Eitel. They consisted of 12 technical men, about 50 men to discharge miscellaneous mechanical and ordnance tasks and about 60 men to handle the various planes and boats assigned to the Section. The 12 technical personnel were bolstered by about 50 independent scientists and technicians, under contract to the private manufacturing firms associated with E-Stelle.

FACILITIES OF TEST STATION (E-STELLE)

The buildings comprising the E-Stelle establishment at Priwall were as follows:

1. Hangar space (Used for offices and small workshops).
2. Workshop and Laboratory buildings.
3. Parachute handling building.
4. Storage sheds.
5. Explosive storage sheds for fuzes, flares, and other miscellaneous small explosive charges.

The vessels assigned to the section were as follows:

1. One ship of about 600 tons, the "Grief".
2. One ship of under 600 tons, the "Norau".
3. Two motor boats.
4. Two crash boats.
5. One medium-sized work boat with mine and cable handling gear.

COORDINATION BETWEEN THE NAVAL MINE AND MINE-SWEEPING GROUP (SVK) AND LUFTWAFFE TEST STATION (E-STELLE).

As a result of the interservice rivalry, the coordination between the two mine development groups, as well as between the various technicians of SVK, was spotty. By this method the separate information of each organization was informally pooled and technical difficulties discussed. When requested, the facilities of SVK were made available to E-Stelle, for whatever purposes desired.

Since the E-Stelle's primary responsibility lay in the testing of items received from manufacturers, they established certain general rules for determining the acceptability of material received. Thus where a lot of one hundred items was received, twenty specimens were selected at random and tested. If 20% or more were unsatisfactory, an additional twenty were selected and tested. If 10% or more of the second lot were defective, the material was rejected. Under special circumstances this method could be altered so that it was more or less stringent, but in every case reports of the tests were forwarded to Air Development (FLE), with explanations and recommendations. In addition, E-7 made spot checks on accepted material stored at the various depots, in order to determine the effect of aging.

Where development work was involved, the first step was the preparation of specifications and requirements by Headquarters with the assistance of Air Development: Mines (FLE-7) and Test Station: Mines (E-7). Thereafter a manufacturer was selected and the project assigned. Members of E-7 technical staff were assigned to provide liaison with the firm. If modifications requested by firms were of minor importance, not affecting the working properties of the article to be made, they were generally allowed. If the firm requested important modifications, because of lack of suitable manufacturing equipment, or if it was unable to carry out important modifications found necessary when the article was put to use, then steps were taken to provide the firm with the requisite equipment. Sometimes there were difficulties which were in reality due only to the policy of the firms in question. In these cases, the advice of outside manufacturing experts, not directly interested in the matter, was requested before any decisions were made. At other times, the firm taking up the manufacture of a particular part would require some patent process of another firm. E-Stelle then arranged to borrow special engi-

ORGANIZATION CHART - SVK

Navy High Command
(OKM: Oberkommandomarine)
|
Mine Command
(SW: Sperrwaffen: Adm. Bachenkohler)
|
Assistant for Mines, etc.
(SWa: Sperrwaffen: Konter Admiral Muller)
|
Mine Inspection
(SI: Sperwaffeninspektion: Vice Admiral Michels)
|

Trials Group (SEK: Sperrwaffen erprobungs kommando: Fgt. Kpt. Broekelmann)	Naval Mine and Mine-Sweeping Group (SVK: Sperrversuchskommando: Kpt. Zur See Bramesfeld)	Mine School	Arsenals
Naval - 1 - Mines (von Linden)　Flotilla (Rodiger) - 2 - Influence Sweeps (Lambrecht) - 3 - Anti-Submarine and depth charges (Davids) - 4 - Mechanical Sweeps (Gemein)	Technical Baudirektor von Ledebur - T -1 Cases Moored Mines (Schuller) - T -2 Sweeps (Behrens) - T -3 Firing Units (Hagemann) - T -4 Cases Aircraft Mines (Kersten) Drafting Rooms Mine Groups Foreign Mines	Shops	Personnel & Administration

Above organizational details supplied by Fgt. Kpt. von Linden.

Figure 1 - SVK Organization.

---

neers a master mechanics familiar with the work, from the second firm for the contractor.

THE EFFECT OF ALLIED BOMBING ON THE MINE PROGRAM

Aside from the creation of transportation bottlenecks, the Allied bombing of Germany had no effect on the production, research, development, or storage of Naval Mines. To safeguard production, the Germans dispersed their contracts among various factories and required that critical parts be manufactured by at least two separate firms. This arrangement proved entirely satisfactory, and no further precautions were taken. On the other hand, the research and development installation of SVK and E-Stelle, although particularly vulnerable to air attack, escaped damage throughout the war solely because the Allied Air Forces chose to neglect them. Similarly the mine depots, with the exception of the one at L'Isle Adam, France, escaped damage throughout the war. The escape of these latter activities was, at least in part, due to their location within heavily wooded areas and to their excellent camouflage, both of which combined to make detection from the air extremely difficult.

GERMAN MINE ORGANIZATION

ORGANIZATION CHART - LUFTWAFFE

Airforce High Command
(OKL: Oberkommando der Luftwaffe)
|
Operations Staff

Chief of Technical Air
Armament
(TLR: Technischen Luft Rustung)

General Staff

Air Development
(FLE: Flugtwicklung)

Procurement

FLE-9 FLE-7 Other Sections
(Torpedoes) (Mines)

LF-12 (Military Control located at Hamburg)

Commander of Test Stations
(KDE: Commander der Erprobungsttellen
Military Control located at Hamburg)

Travemunde

Other Test Stations
(Erprobungsttellen)

E-5 E-7 E-9
(Mines) (Torpedoes)

Hauptmann Eitel

Group 1 Group 2 Group 3
(St. Ing. Spieler) (Hpt. Ing. Kern) (St. Ing. Doorman)

Figure 2 - Luftwaffe Organization.

Chapter 3

CRITIQUE

Throughout a large part of the war, Germany possessed a high degree of technical advancement in the field of marine mines, but never fully exploited it. A number of factors prevented the German mines from reaching their point of potential destructiveness against the Allies. At first the Navy placed little importance upon them; subsequent inter-service rivalries seriously impaired their effectiveness; and, in the latter stages of the war, shortages curtailed operations. In addition, miscalculations at several points marred the program.

At the beginning of the war, the German Navy emphasized guns and torpedoes. Apart from a small group of specialists, it was not interested in mines. Even the specialists believed that while mines, intelligently used, could be powerful weapons, they were very likely to be discredited by injudicious use. This liability to fortuitous rise and fall in the stock may well have contributed to deprive mine development of the consistent direction and drive to be seen in German torpedo development.

Mining suffered consistently from its subordinate standing. No one in the Navy held operational control over mining; no one in the Mining Command possessed sufficient drive and grasp to present the case for mining with enough force before the High Command. As a result, the direction of mine warfare failed to rise above its second-rate position.

The weakness of the Mining Command was readily apparent in operations. The decentralization of control over operations and operation policy contrasted markedly with the highly centralized control the Mining Command held over materials. The German Navy had no specially built, high-speed minelaying vessels capable of large-quantity plants. Although Schnell boats, submarines, destroyers, cruisers, and certain merchantmen were fitted for minelaying, none of these vessels was ever available in sufficient number of combination to meet the strategic requirements of the mine group.

Because mine priority was disproportionately lower than that of torpedoes, too few submarines were assigned to minelaying in American waters. As a result, there were no effective minefields in the western Atlantic to disrupt coastwise shipping and convoys to Europe.

The field commanders persisted in laying only those mines designed to sink merchant tonnage, since such sinking made better press-release material and created higher morale than did the sinking of small minesweeping vessels. Pressure from the High Command finally resulted in a change of policy, but by then it was too late.

The greatest weakness in the mining program was the lack of cooperation between the Navy and the Luftwaffe, and on a lesser scale between the Navy and the Army. The Luftwaffe insisted upon its own independence, and the Naval Mining Section (1 Skl) had no jurisdiction over its minelaying activities. The Navy maintained that every minelaying operation was a naval operation. Accordingly they tried to influence policy, although they could not exercise control; but even in this they had little success.

After the collapse of France, when the Navy came to the view that an effective sea blockade of England would bring her to her knees, the Luftwaffe continued to use bombs, the results of which were tangible and of greater propaganda value. Admiral Muller declared, "Goering was interested in showing Hitler and the German people pictures of bombed and burning English cities, and was not content with the invisible and often immeasurable results of Naval Mine Warfare."

At the same time, the Luftwaffe preempted much of the Navy's jurisdiction over both operation and production. The pre-war plan had been for aircraft minelaying to be confined to estuaries and such coastal waters as could not be reached by surface craft or submarines. One Luftwaffe formation based in northwest Germany and cooperating with the Fleet was to carry out all aircraft laying. The exclusion of German naval forces from British waters led the Luftwaffe to extend its area of operation, and the plan quickly broke down. In its minelaying it so completely ignored the Navy that it prematurely laid two new types of mines before they were ready in large numbers, and thus helped destroy their surprise effect.

Interservice politics undoubtedly had their part in the decision of the Luftwaffe, taken about the beginning of the war, to develop its own bomb-mine to replace the Navy parachute mine. The Luftwaffe placed an order for the first bomb-mine without any knowledge of the principles of that type of mining, and with the sole specification that it be of the same size and shape as a bomb. The tactical considerations behind the decision were no doubt sound enough, but, at the time, the Luftwaffe had no technical staff of its own which was sufficiently versed in mining problems to obtain

RESTRICTED

a balanced solution.

When the Luftwaffe undertook mining developments, available experimental and testing facilities were very small and temporarily makeshift. The only equipment available was some generally used for torpedo work at Travemunde. Adjoining was an airfield originally used for experimental work with seaplanes. Subsequently a testing station for mines was constructed, and changes made in personnel.

At the time developments began at Travemunde, the staff had no experience with mines. The original suggestion to transfer trained personnel from the Navy was rejected, and it took some time to train the necessary staff and initiate testing and development on the requisite scale. Finally, after much argument, an officer with mine experience was assigned to the station in 1943. Later a strong technical staff and considerable development resources, mainly in industry, were built up.

Ultimately the Luftwaffe put much effort into the design of firing systems and corresponding modifications of the BM(bomb mine). Nevertheless, the BM was more restricted in its condition of drop than the Navy LM (parachute mine). With the exception of the pressure unit, which was a special case, the Navy had already produced all essentially new firing systems, and installed them in the LM.

When production work started on the pressure mine, the Luftwaffe and the Navy disagreed on principles and design. As a result, both services manufactured their own versions. Later, when the unit was ready for laying, the services again disagreed on its use in combination with other units. The Navy insisted on combining it with the magnetic unit, while the Luftwaffe preferred its use with an acoustic unit.

The Navy seldom attended trials of new devices at Travemunde, but it did receive completed specimens, sent for information and suggestions.

While the war lasted, the diversion of effort involved in the dual development continued. Luftwaffe Colonel Rommel complained that the naval system was far too rigid to get results quickly. Since the Luftwaffe did not get them any quicker, the view of German naval officers that the separation was thoroughly undesirable, seems to be sound.

To a lesser extent, the same sort of division existed between the Navy and the Army. Although the Navy was responsible for harbor security and control mines, they had no cognizance of any over-all beach defense mining plans developed by the Army. Coastal defenses and anti-invasion matters were under the jurisdiction of area commanders. Mines intended for use against river shipping, against bridges, and for anti-invasion purposes were extremely simple to manufacture, and therefore produced and procured locally by area commanders. The Navy was seldom informed about such improvisations, and believed that in closer cooperation with the Army better results might have been obtained.

TECHNICAL DEVELOPMENT

Despite the fact that the first German magnetic mine unit was ready for operational use in 1925, at the outbreak of the war in September 1939, insufficient stocks of magnetic mines were on hand to wage an all-out and effective mining campaign against England and her Allies. This shortage existed because the war came at a time when the Navy was still engaged in improving existing operational models and had not, as yet, gone into mass production. The existing stocks of magnetic mines were very small. They consisted of approximately six hundred LMA's and LMB's and several hundred RMA's and RMB's.

The German technical development of mine firing systems fell into three stages. The first, the period from October 1939 to the Summer of 1941, was one of significant innovations. It saw the introduction of the magnetic mine with its successive modifications, followed by the rush development of the audio frequency acoustic and the combined magnetic-acoustic mines.

The German Navy expected the magnetic mine to suffice for the war, but within a year the British were applying successful countermeasures. The Navy placed the blame for this upon the Luftwaffe, which in 1939 laid the mines in the Thames estuary before enough were available for a heavy attack. The Navy felt that these were far more likely to be recovered than ship-laid or submarine-laid mines, which might safely be laid in small numbers.

Even before the British negated the effect of the magnetic mine, the German Navy began the rush development of the audio frequency acoustic and combined magnetic-acoustic mines. As a result, they were actually producing acoustic mines in small numbers within three months after the outbreak of the war. Quickly they overcame the operational limitations of the first models, and proceeded in 1941 to develop a combined magnetic-acoustic mine. Once more the Luftwaffe stultified the effect through premature laying.

By the end of the period the effectiveness of the magnetic mine had seriously diminished. The Germans began, as the basis for a policy of technical surprise, a systematic study of ships' influence fields.

The second period, from the introduction of the magnetic-acoustic mine in the Summer of 1941, to the end of 1943, was one of steady research but no essential novelties. In April 1942, the German Navy had approximately 50,000 mines of all types ready for operational use. Through 1942 the monthly demand for mines was extremely low and at a constant level. Subsequently, as Allied action became more aggressive, especially in the Mediterranean, the demand accelerated. Ironically, now that no

GERMAN UNDERWATER ORDNANCE—MINES

new weapons were coming out and successful countermeasures for the old ones were in operation, the Luftwaffe took mining more seriously and put forward its biggest minelaying effort.

During the third period, from the beginning of 1943 to the end of the war, the design work of the previous two and a half years bore fruit. The pressure, the low-frequency acoustic, and variants of the audio frequency circuits came into service, and the Navy had under development a wide variety of other weapons. From 1943 on, the Allies had studied the technical problems involved in countering many of these weapons; but their use would, none the less, have been very unwelcome.

The Germans had developed not only new types but also new techniques. Since no single mine is insurmountable, the weapon was the minefield, not the individual mine. Mine warfare operated on the principle of statistics. At the same time that German mines were becoming increasingly complicated in firing principles, first the Luftwaffe and subsequently the Navy, late in the war, arrived at the general policy of laying mixed fields. These greatly increased the problems of sweeping.

The demand for mines reached its peak in the invasion year, 1944. By then the Naval High Command appreciated the value of mine warfare, but was forced to cut orders to conform with the maximum possible production. At the same time, the laying capacity of the Luftwaffe declined heavily as a result of the German reverses on all fronts. This took the sting out of the new armory of mines. E-boats and other naval craft took over the mining offensive in the last stages of the war, but their scale of operations was necessarily small.

The Allies were fortunate that, very much as in the case of German submarines and torpedoes, the bomber position was not in phase with the weapon position. If the Luftwaffe had regained its offensive power, the Allied mine defense would have deteriorated.

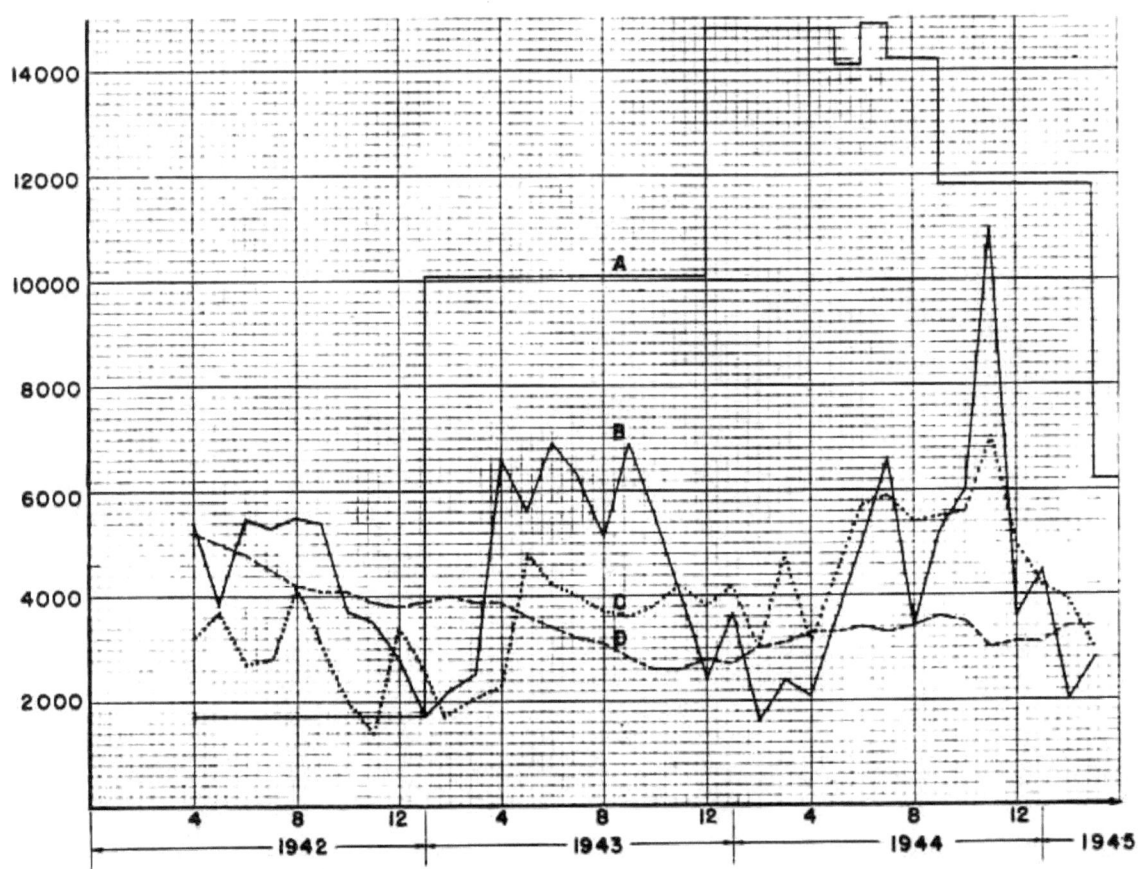

A- MILITARY DEMAND
B- MINES LAID
C- AMOUNT RECEIVED
D- SUPPLIES ON HAND (x10)

Figure 3 - Graph of Mines Laid.

CRITIQUE

MINES LAID

Year	Month	BMC	FMB	FMC	UMA	UMB	OMA/I	BMC/S	BMS	BMA	BMH	TMB I/II	TMB III	TMC I	TMC II	EMB	IMA/S	LMB/S	BMA II	BHF	LMF/S	SMA	Total
1943	10	1140	64	305	6	1324			4			60		8				403		497	50	95	3956
	11	328		53		679			3	37	166	178	80					94		907		27	2515
	12	1536			330	574			5	37	47	30					150	937		192			3888
1944	1	885	11			284				3	40		129					180		247	28		1560
	2	1184			44	549		26				117						191		247			2358
	3	1678				126				22	24	8						63		246			2167
	4	1074			186	499		21	5									752		1095		30	3602
	5	2008	16		73	1341		63		174	140	45					194	642		66	36		4854
	6	1429		54	34	1723		364		1	82	70					310	1846		253	200	28	6572
	7	73			234	937		156			27	44						1117	7	1015		1	3612
	8	1541			850	838		180	15	163	129	99		1	60	1	70	874		158	130		5107
	9	767			510	506		150	26	57	199	433		98			364	2183		429	100	51	5972
	10	3485		8	830	2105			19	27	598	57		20		16	77	1917	1	1697		173	11016
	11	1514			86	593		1	5	24		58	12	47	15	10		1056		148		66	3635
	12	1135			84	694				201		158				1041		950		245			4508
1945	1	830				329											29	194	94	557			2033
	2	664			1	559	100		1					40	6	1002	1	367	91	64			2895
	3	1269			497	500				50		19	90	155	12	14		532					3136

TOTAL NUMBER OF MINES LAID PRIOR TO OCTOBER, 1943

Year	Month	Total	Year	Month	Total	Year	Month	Total
1942	3	5400	1942	9	3700	1943	3	6600
	4	3800		10	3500		4	5600
	5	5500		11	2800		5	6800
	6	5300		12	1700		6	6300
	7	6000	1943	1	2300		7	5200
	8	5400		2	2500		8	5800
							9	5400

Figure 4 - Mines Laid.

Chapter 4

CONTACT AND MOORED INFLUENCE MINES

THE GERMAN EM (EINHEITSMINE) MINE SERIES)

The German Einheits Minen (EM) mine series consisted of 13 different types of sea mines. These types were designated EMA, EMB, EMC, EMD, EME, EMF, EMG, EMH, EMI, EMK, EMR, EMS, and EMU. With the exception of the EMS, which was a drifting mine, all of the series were moored mines laid by surface and/or underwater craft.

The EMH, EMI, EMK and EMU either were abandoned in the developmental stage or were incomplete at the close of the war, and no specimens of any of these types or documents relating thereto were found. Consequently, the information contained herein on such mines is based solely on statements made by German prisoners of war and should be treated accordingly.

The EMA Mine. The EMA was developed during World War I and was the first German mine with a chemical-horn firing system. Accordingly, to differentiate it from the pendulum-type mines then in use, it was designated Elektrische Mine type A. Its production was discontinued at the close of the war. Its appearance in World War II is accounted for by the fact that stocks remained on hand in 1939.

The mine existed in two models, one for laying by surface craft and the other by submarine. Only the surface craft type was laid in World War II, since the other type required specially fitted submarines which were not available.

Description of Case

Shape	Two hemispheres, joined by a 12-in. cylindrical mid-section.
Material	Steel
Diameter	34 in.
Length	46 in.
Charge	330 lb. block-fitted hexanite

Description of External Fittings

Horns	Five: one in center of upper hemisphere; four, equally spaced, around upper hemisphere
Arming switch and booster release	On mid-section, secured by keep ring
Detonator carrier mounting	In bottom center of case
Mooring bracket and white metal mooring switch	Bolted to two lugs on lower hemisphere

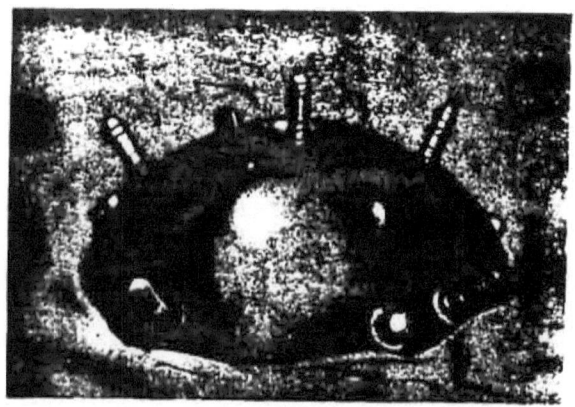

Figure 5 - EMA Mine Afloat

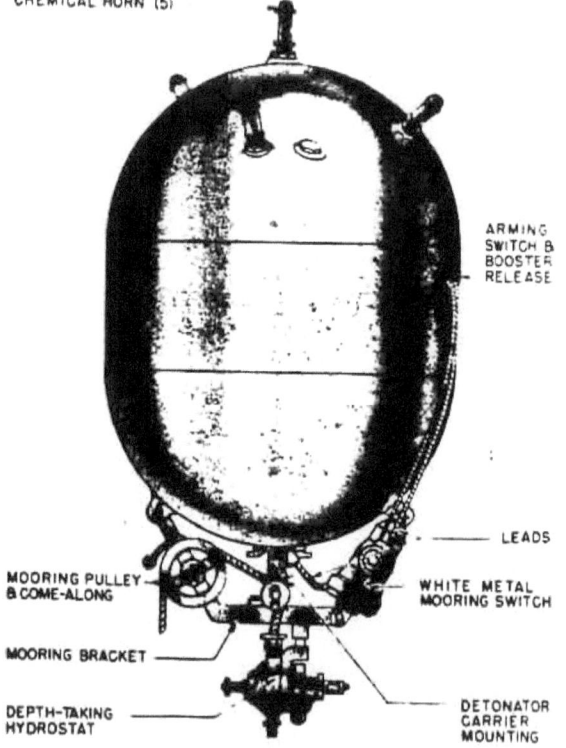

Figure 6 - EMA/EMB Mine

Mooring pulley and   Attached to extension
  "come-along"   of mooring bracket

Depth taking   Bolted to extension
hydrostat   on mooring bracket

Two pair of electrical leads extend from the white metal mooring switch, one set to the detonator carrier, the other to the arming switch.

Operation. Mine takes depth by hydrostat. Separation of the anchor and case withdraws a safety pin from the arming switch and booster release, making the circuit from the horn batteries to the detonator and allowing the booster to drop over the detonator. Mooring tension extends the spindle of the white metal mooring switch, arming the circuit of the internal horn to arm the mine.

The EMB Mine. This mine is identical to the EMA except for the weight of charge, which differs as follows:

1. EMA   330 lb.
2. EMB   485 lb.

It is a moored contact type fitted to take seven chemical horns, and was designed for use against surface and underwater craft.

Figure 8 - EMC Mine Afloat

EMC I - EMC II - EMC II (Upper Antenna)

General

Moored, contact, chemical-horn mine, laid by surface craft.

Offensive or defensive mine, for use in maximum depth of water of 1700 feet.

Maximum depth of case when moored is 245 feet.

Description of Case

Shape	Two hemispheres, joined by a 2 cylindrical midsection
Material	Steel
Diameter	46 in.
Length	48.5 in.
Charge	660 lb. block-fitted hexanite

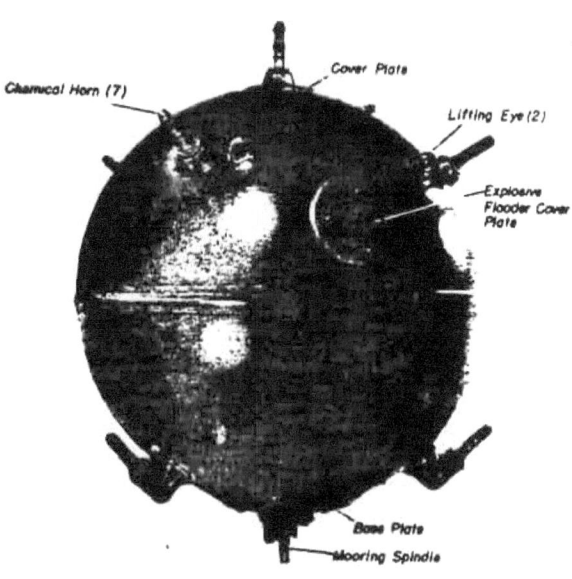

Figure 7 - EMC I Mine

The EMC I Mine. The original EMC mine was completed in 1924. (figure 7). It utilized a bronze base plate and an eighth horn (KE) which was placed on top of the mooring safety switch. In 1936 the mine was improved. The eighth horn was moved to the side of the mooring switch; a soluble plug was substituted for a dash-pot type of delay system; fittings were provided for antennae and scuttling devices; and the base plate parts were redesigned to give greater life and watertightness. The improved model was designated EMC II and the original type redesignated EMC I.

Description of External Fittings

Horns	Seven: one in center of cover plate; four equally spaced around upper hemisphere, 22 in. from center; two on brackets, 39 in. apart, 17 in. from center of lower hemisphere
Cover plate	7½ in. diameter, in center of upper hemisphere, flush type, secured by 10 bolts

Figure 9 - EMC II Mine with Rubber Snag Line

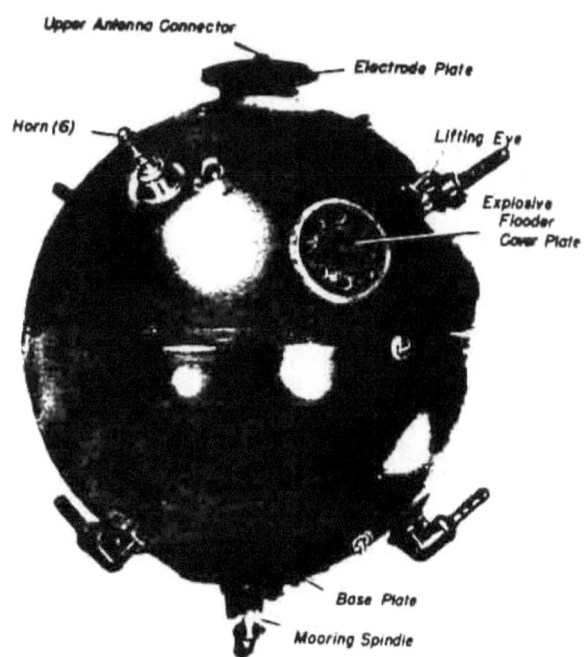

Figure 10 - EMC II Mine

Base plate   Standard type EMC II

Lifting eyes   Two, 19 in. apart, 22 in. from center of upper hemisphere

Operation. Mine takes depth by plummet. Mooring tension pulls out the mooring spindle, closing the mooring safety switch, tripping the booster release lever and the mine is armed.

Standard chemical-horn firing.

The only self-disarming device is the mooring safety switch which is designed to disarm the mine by opening the firing circuit upon release of mooring tension.

EMC II Mine. The EMC II existed in the six types shown in figure 49a. The general characteristics of each of the various types are as follows:

EMC II with Upper and Lower Antenna

    Upper antenna   130 ft.
    Lower antenna   100 ft.
    Depth setting   3 to 190 ft.
    Max. depth of case 245 ft.

EMC II with Tombac Tubing

    Tombac tubing is an anti-sweep cable fitted over the mooring cable

    Tombac Tubing   100 ft.
    Depth Setting   3 to 190 ft.

EMC II with Lower Antenna

    Lower antenna   100 ft.
    Depth setting   3 to 190 ft.

EMC II with Cork-Floated Upper Antenna

    Upper antenna   65 ft.
    Depth setting   3 to 190 ft.

EMC II with Chain Mooring

    Chain mooring   20 ft. of 5/8-in. chain
    Depth setting   3 to 190 ft.

EMC II with Chain Mooring and Cork-Floated Snag Line

    Chain mooring   20 ft. of 5/8 in.
    Snag line   80 ft. with cork floats
    Depth setting   40 ft.

In 1940 the eighth horn was removed from all German base plates, because experience had shown that this device was often actuated in heavy seas. The upper antenna was abandoned in 1941 because of the excessive numbers that broke loose in rough waters; and, in 1943, the lower antenna was abandoned because of the copper shortage then prevalent.

Experiments were conducted to obtain delayed rising of the mine. A fifty-foot bight of the mooring cable was flaked on the top of the anchor and kept in a locked position by a six-day clock. In operation the mine would plant at its set depth and, when the clock had run

CONTACT AND MOORED INFLUENCE MINES

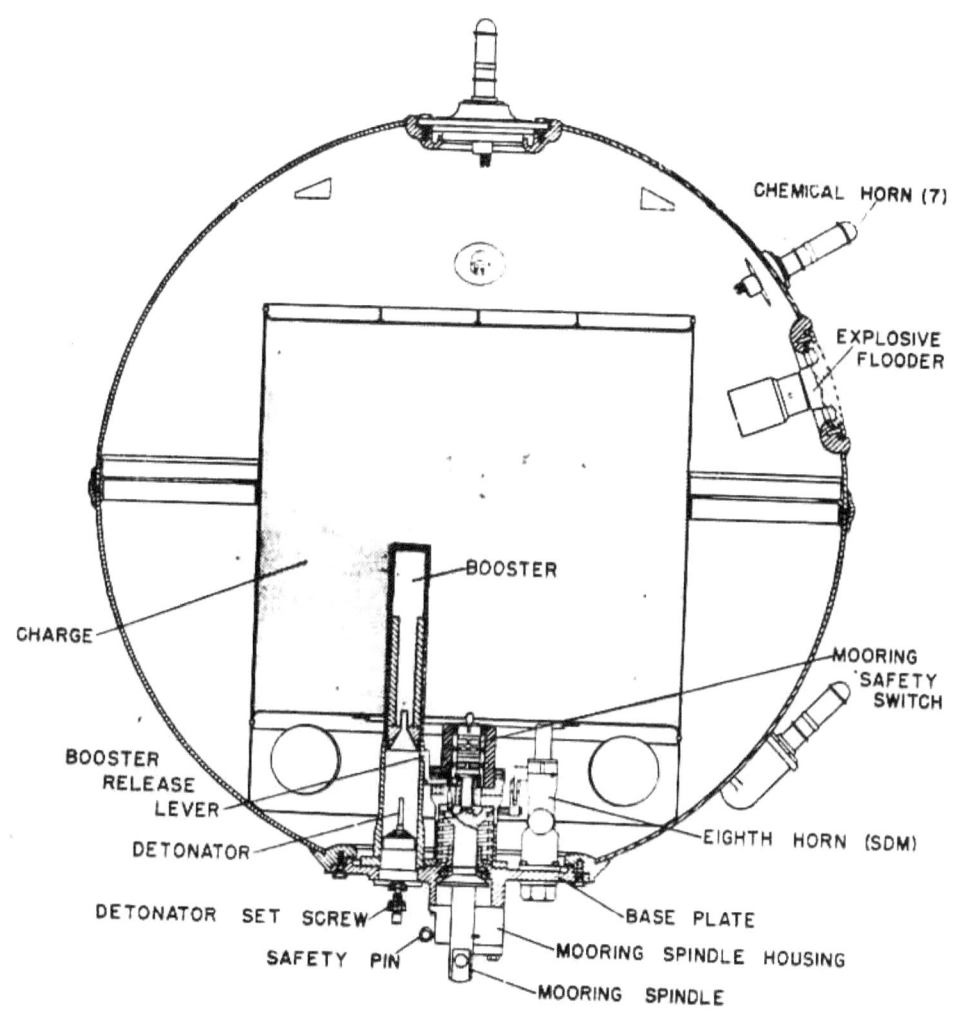

Figure 11 - EMC II Mine - Cross Section

off, rise 50 feet. This idea proved unworkable because of the excessive strain created on release of the bight.

In 1944, because of a critical shortage of lead, steel horns were developed and substituted for the standard Hertz lead-acid horn. The steel horns were so constructed that the metal and welds would not part if bent to an angle of 90 degrees. A pull of approximately 130 pounds is sufficient to bend the horn and break the inner vial. Although the lead horns were considered superior, the steel type proved satisfactory operationally.

Early in 1944, the German Navy experimented with a 32-second clockwork release device to replace the standard dash-pot plummet delay. This clockwork was standard Luftwaffe equipment used to obtain delays in the opening of cargo parachutes, etc. This device was simple and easy to produce. It operated as follows:

The time delay desired, up to 32 seconds,

was selected by turning a dial on the face of the clock. The clock was simultaneously wound and started by pulling a wire lanyard at the base of the device. At the opposite end, another wire was run off: the wire lanyard was snapped in by the spring-loaded clock drum and the safety pin withdrawn.

When, with relatively minor casualties, the Allies succeeded in penetrating the field of moored contact mines laid off Salerno, the German Mining Command suspected that ASDIC was being used to locate the moored mines. Accordingly, they gave some thought to the development of a mine case that would resist location by such methods. They sought to attain this end by coating mine cases with rubber and using special type paints. To this extent, their efforts paralleled those of the German Submarine Command, which sought to apply anti-detection methods to submarines. However, the foregoing methods proved unsatisfactory, and it was finally decided that the best anti-detection type mine would be one employing

RESTRICTED

GERMAN UNDERWATER ORDNANCE—MINES

Figure 12 - EMC II Mine with 80-ft. Rubber Snag Line

Figure 13 - EMC II Mine with 20-ft. Mooring Chain

an all sponge-rubber case. Limited experimentation was commenced to determine the response of various types of synthetic sponge-rubber to ASDIC. The end of the war caught these experiments in their early stage. Consequently, no mine cases of this type were actually built.

Both models of this mine were laid operationally. (A field of EMC I mines was laid in the South Pacific, some of which were recovered by U. S. Navy Mine Disposal Personnel.)

The EMD Mine. The first EMD mine was ready for operational use in 1924. It was a moored contact type fitted for five chemical horns, and was designed for use against surface craft only; consequently, it had no lower horns. Except for the absence of such horns and its smaller size, the EMD is practically identical to the EMC. (Both mine types use the same base plates, anchors, and accessories). In 1936 it was improved along the same lines as the EMC, the new model being designated EMD II and the original type EMD I. A small cover plate, 6.5 inches in diameter, equidistant from the lifting eyes and 25 inches from the center of the upper hemisphere, was added to accommodate an 80-day clock and flooder. Later, an electrode plate mounted on a plastic cover and

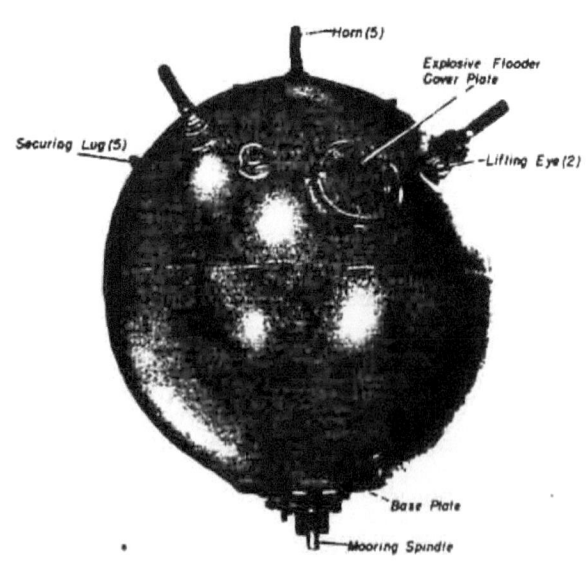

Figure 14 - EMD I Mine

CONTACT AND MOORED INFLUENCE MINES

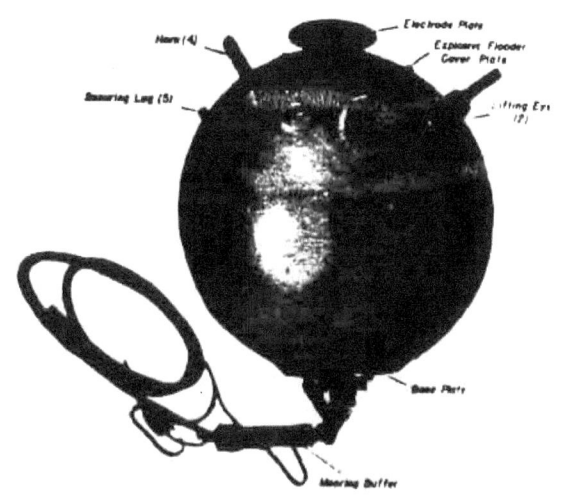

Figure 15 - EMD II Mine - Upper-Lower Antenna

placed in the center of the upper hemisphere was added as an antenna connector.

The manufacture of EMD II was discontinued in the early part of World War II to permit greater production of the EMC II, which was considered more suitable. Existing stocks of EMD I and EMD II were laid operationally.

General

Moored, contact, chemical-horn mine, laid by surface craft

Offensive or defensive mine, for use in maximum depth of water of 1000 feet.

Description of Case

Shape	Spherical
Material	Steel
Diameter	40 in.
Charge	330 lb. block fitted hexanite

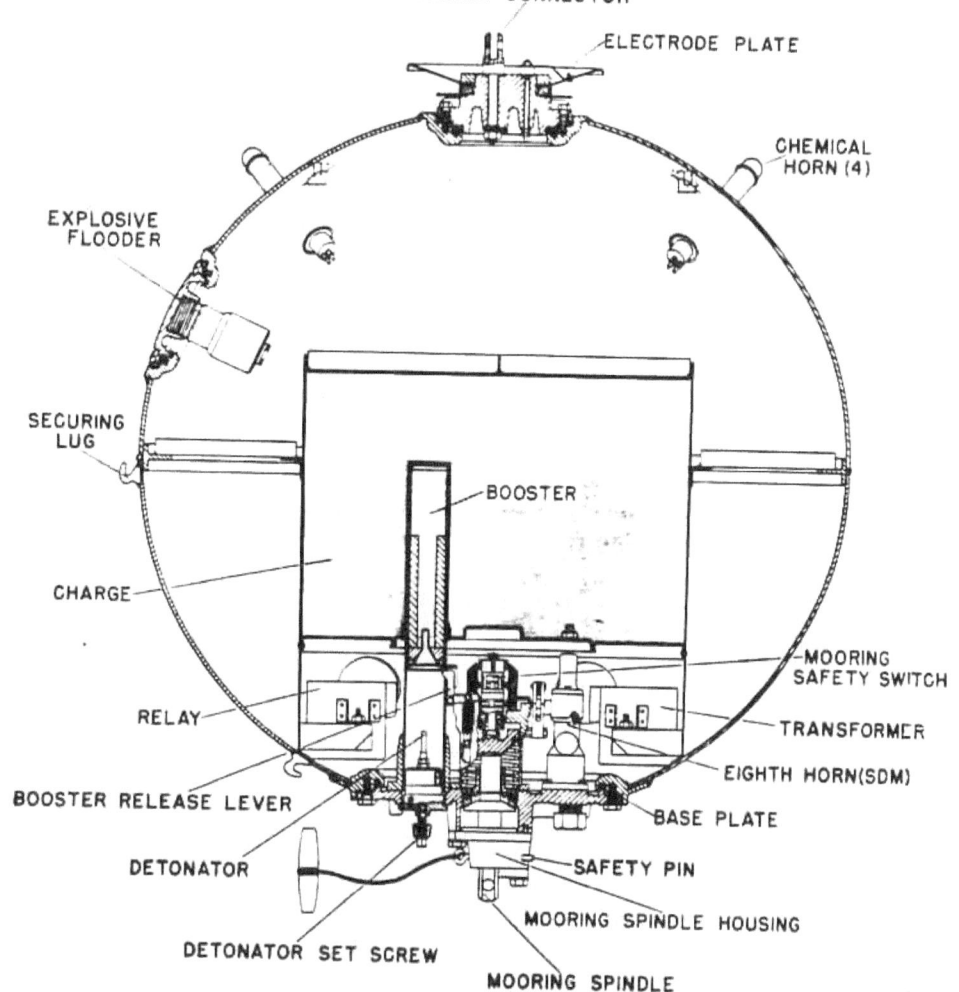

Figure 16 - EMD II Mine - Cross Section

RESTRICTED

GERMAN UNDERWATER ORDNANCE—MINES

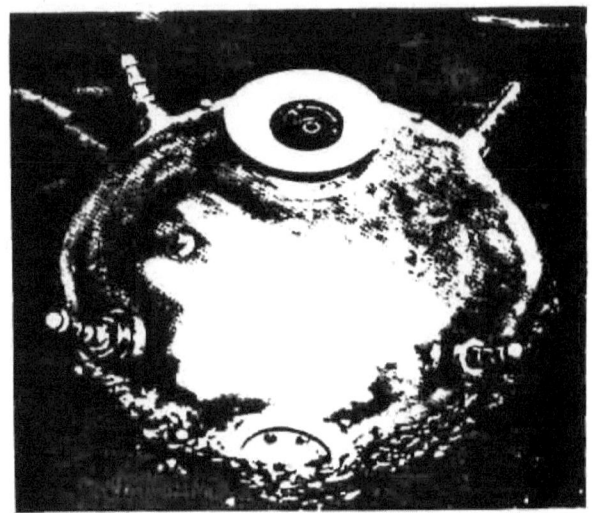

Figure 17 - EMD II Mine Afloat

Description of External Fittings

Horns	Five; one in center of cover plate; four equally spaced around upper hemisphere, 20-in. from center
Cover plate	7.5-in. diameter, in center of upper hemisphere flush type, secured by 10 bolts
Base plate	Standard Type EMC II
Lifting eyes	Two, 16.5 in. apart, 22.5 in. from center of upper hemisphere
Securing lugs	Five; one 22.5 in. from center of upper hemisphere; one 31 in. from center of lower hemisphere; three, staggered, 12 in. from center of lower hemisphere

Operation. Mine takes depth by plummet. Mooring tension pulls out the mooring spindle, which closes the mooring safety switch, trips the booster release lever, and arms the mine.

Standard chemical-horn firing.

The only self-disarming device is the mooring safety switch which is designed to disarm the mine by opening the firing circuit upon release of mooring tension.

The EME Mine. This mine was a moored contact type bought from a British firm and designated "Elektrische Mine Englische" (EME). It was used solely for experimentation. No details of this mine are available.

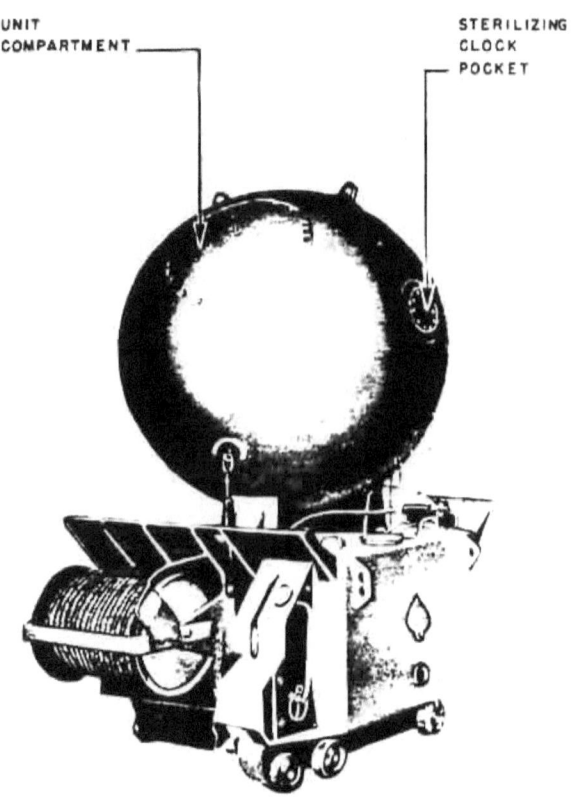

Figure 18 - EMF II Mine

The EMF Mine. The EMF was the first moored influence mine developed by the Germans. Its design was undertaken in 1928 and completed in 1931. In 1936 the base plate was revised and the mine put into production. By 1939 it was ready for operational use, but the magnetic unit in existence proved unsatisfactory. In 1941 the M 3 had been perfected and adapted for use with the EMF. At about the same time unsuccessful attempts were made to fit this mine with an acoustic unit known as A 3. The mine was laid operationally only with the M 3 unit. It was contemplated that the following influence unit should be fitted to the mine:

1. M 4
2. AA 4 (Unit abandoned in 1944)
3. A 7
4. AE 1

The EMF used the EMC anchor and was designed for surface laying only (figure 18). Its larger counterparts, the SMA, SMB, and SMC were, on the other hand, designed for submarine laying only.

The first model was a spherical case consisting of two hemispheres welded together. Its flooder plate was located $26\frac{1}{2}$ inches from the center of the upper hemisphere; however, there were no provisions for an 80-day clock. In the final model the case was improved in

CONTACT AND MOORED INFLUENCE MINES

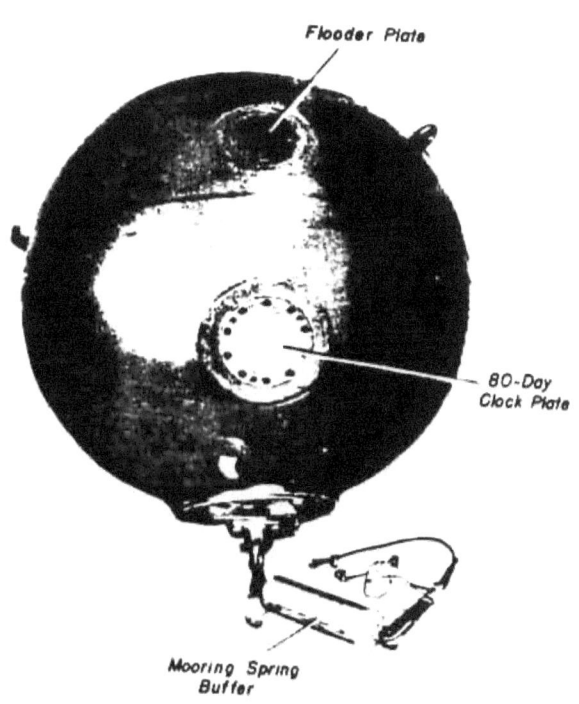

Figure 19 - EMF Mine

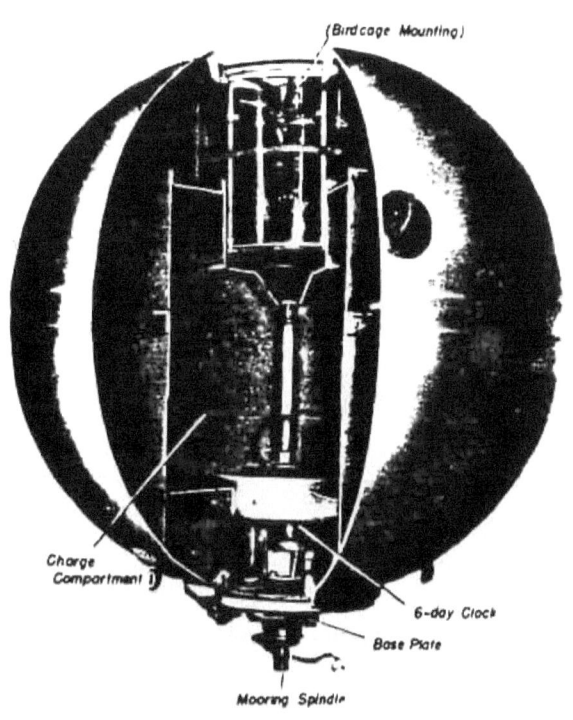

Figure 20 - EMF Mine with M 3 Unit

construction; a soluble plug and disarming switch were added; and the depth taking mechanism was improved.

Description of Case

Shape	Two hemispheres, joined by a cylindrical mid-section
Diameter	45 in.
Length	50 in.
Charge	750 lb. block-fitted hexanite

Description of External Fittings

Lifting eyes	Two 60° apart on upper hemisphere, 20½ in. from center
Anchor-securing lugs	Three, hook shaped: two on lower hemisphere, 160° apart, 11¼ in. from center; one on upper hemisphere, 28½ in. from center
Flooder plate	6-in. diameter, on upper hemisphere, 28½ in. from center, secured by 10 bolts
80-day clock cover plate	8-in. diameter, on lower hemisphere, in line with flooder plate, 23½ in. from center, secured by 10 bolts

Operation. The mine takes depth by plummet. Mooring tension pulls out the mooring spindle, tripping the booster release lever and releasing the locking balls from the clockwork spindle. Water pressure depresses the clock spindle at a depth of 15 feet, starting the clock. The clock runs off its delay period, and the unit starts its testing cycle. If the mine does not orient itself properly after a pre-set time of up to 12 hours, a scuttling charge will fire to sink the mine.

The only self-disarming device is the 80-day clock, which is designed to scuttle the mine if the clock stops at any time prior to completion of its set period or upon completion of its set period.

The operational characteristics of this mine are as follows:

Laying heights and speeds	13 ft. - 25 knots 16 ft. - 18 knots
Laying depths	325 ft. - with cable ¼ in. 650 ft. - with cable 7/16 in. 985 ft. - with cable 3/8 in. 1640 ft. - with cable 5/16 in.

RESTRICTED

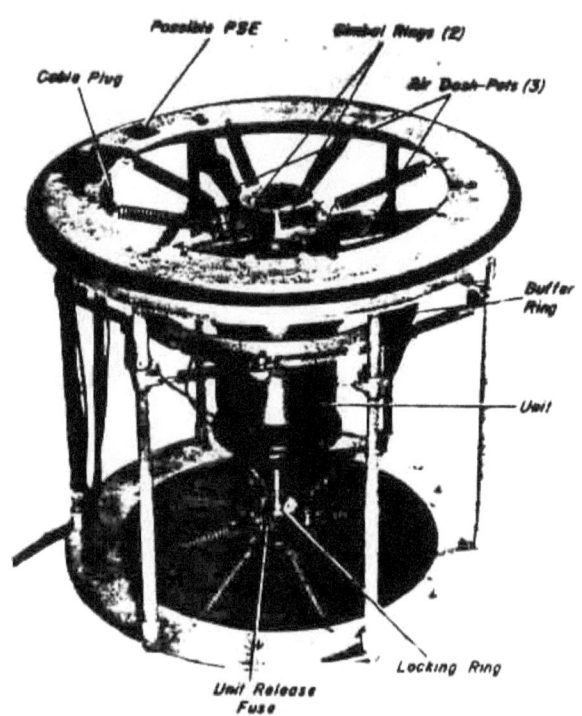

Figure 21 - EMF Mine - Birdcage Suspension for M 3 Unit

Minimum laying depths	130 ft. (80 + 50)
Minimum and maximum case depths	50 to 115 ft.

In 1944 experiments were conducted with an EMF case made of a plastic material called "Eternit". This model was known as EMF (Et) and consisted of two Eternit hemispheres bolted together to form a sphere similar in dimensions and fittings to the normal EMF. It was undergoing tests at the close of the war.

The EMG Mine. The EMG was a moored, contact, constant-depth mine assembly designed for defense against small surface craft such as torpedo boats.

The assembly was designed in 1940 to protect German shipping in the English channel from attack by British torpedo boats and other similar craft. (The original plan contemplated that this assembly would be employed to protect the flanks of German shipping lanes established in an invasion of England.) The assembly was used operationally until 1943, at which time it was abandoned in favor of the UMA/K and OMA type mines.

The EMG assembly consisted of a ballasted EMC mine case with the lower horns blanked off, an EMC anchor, a float, and a weight arranged for constant depth. The assembly was so designed that it maintained a constant depth of eight feet, regardless of the stage of the tide.

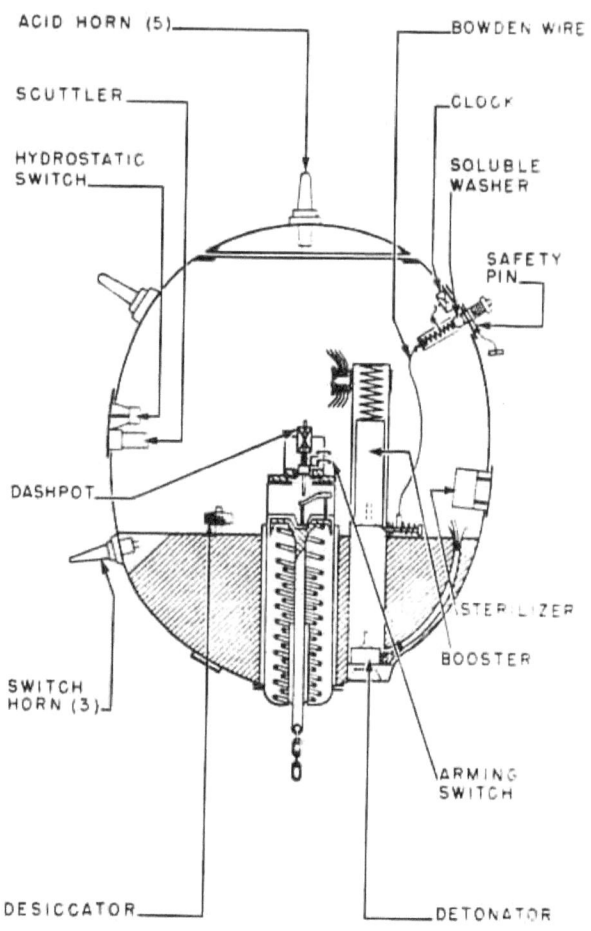

Figure 22 - EMU Mine

This assembly could be laid in depths ranging from 30 to 190 feet; but, by lengthening the mooring cable between the anchor and the weight to 820 feet, the assembly could be used in greater depths.

Since the EMG float rode slightly above the water-surface, mine fields utilizing this assembly were easily detected and avoided. To make this apparent disadvantage inure to their own benefit, the Germans developed a dummy EMG assembly which consisted merely of the normal float and anchor, and a 325-foot length of mooring cable. These dummies were laid in separate fields or together with EMG's. They were designated "Simulacker fur EMG."

The EMH Mine. In 1942 the Germans were still seeking to develop an acoustic unit for use in moored mines. Since the aluminum EMF was expensive to build, it was decided to design a cheaper moored mine case of sheet-iron construction, to be known as the EMH. However, since the design of an acoustic unit for moored mines was progressing very slowly, the development of the EMH was discontinued.

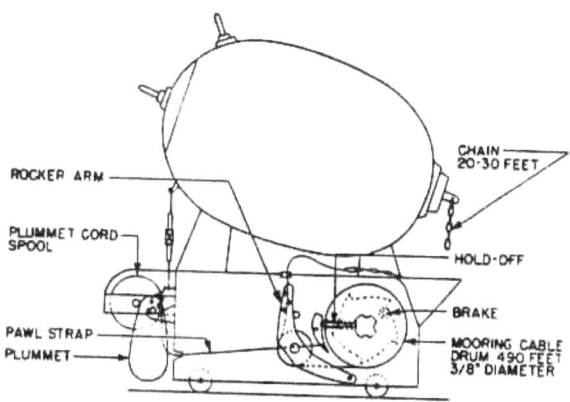

Figure 23 - EMU Mine and Anchor

It was intended that the EMH be laid by surface craft, be the same size as the EMC, use an EMC anchor, and be capable of the same depth settings as the EMC.

The EMI Mine. In 1940 the Germans undertook the development of a mine case that would house an induction-type magnetic unit. To reduce development work, they intended to utilize a suitably modified EMC case, which was to be known as EMI. After preliminary development work had been done, the mine was dropped for the following reasons:

1. Shortages of nickel and copper prohibited the large scale use of naval induction units.

2. Induction units had not been sufficiently developed to permit their use in moored mines.

The EMI was the only attempt by either the German Navy or the Luftwaffe to develop a moored induction-type mine.

The EMK and EMU Mines. The development of the EMK was undertaken in 1940, the mine being intended for use as a moored contact and/or influence-type mine. In 1944 its development, which was still incomplete, was discontinued in favor of the smaller EMU. Since the EMK and EMU were identical, except for size (EMK 44 inches in diameter and EMU 40 inches), they are discussed together in the following paragraphs.

The EMK was dropped in 1944 in favor of the EMU because of a shortage of explosive. (The EMK was designed for a charge of 660 lb., the EMU for 220 lb.) Since these mines were radical departures from previous German types (figures 22, 23), their development progressed slowly. Consequently, at the end of the war neither the EMK nor its successor, the EMU, was completed.

By 1940 the Germans had realized that their standard-type base plates for moored mines had two serious shortcomings:

1. In deep water, hydrostatic pressure sometimes prevented arming by counterbalancing the pull of the mooring cable.

2. In shallow water, rough seas caused excessive arming and disarming, and frequently wore out the spindle-mechanism membrane.

To cure these defects an entirely new type of base plate was designed. (figure 22). This base plate utilized an inverted spindle action, so that water pressure and mooring-cable tension combined to arm the mine. In depths over 30 feet, hydrostatic pressure alone would maintain the mine in an armed position, regardless of vertical motion of the mine case. In water depths of less than 30 feet, disarming due to vertical motion of the case was prevented by a dash-pot mechanism which maintained arming of the mine for 60 seconds after release of tension on the mooring spindle.

The method of placing the main charge in the EMK and EMU differs from normal German practice. Whereas all other German moored mines employ a charge container, the EMK and EMU were designed to house the charge on the bottom of the mine case. The Germans felt that loading the mines in this manner would give the mine a greater lethal range and permit better mine orientation.

The anchor of the EMK and EMU mines was also of new design. Its most noteworthy feature was the fact that it was so designed as to permit depth setting either by plummet or by hydrostat (figure 23). The plummet-line drum was designed to accommodate 100 feet of 3/16-inch cable, and the mooring-cable drum 500 feet of 7/16-inch cable. An 18-30-foot length of chain was to be used between mooring spindle and mooring cable. Because of its departure from previous German types, the development of this anchor progressed slowly and was not completed at the end of the war.

EMR and EMR/K Mines. These mines were actually sweep-obstructors utilizing an EMC mine case moored to an EMC anchor by a single or double length of 5/8-inch standard chain.

The EMS Mine. The EMS (Sehrohrtreibmine S) is a drifting decoy or anti-pursuit type of mine. It existed in three forms, which were designated EMS I, EMS II, and EMS III.

The three types of this mine employed the same mine case; they differed only in the method of flotation.

The characteristics of the mine case are as follows:

Method of firing	5 sensitive switch horns
Weight of charge	24-30 lb.
Weight of case without flotation gear	100 lb.
Height of case	21 in.
Diameter of case	13 in.
Diameter of case including horn bosses	18.5 in.

GERMAN UNDERWATER ORDNANCE—MINES

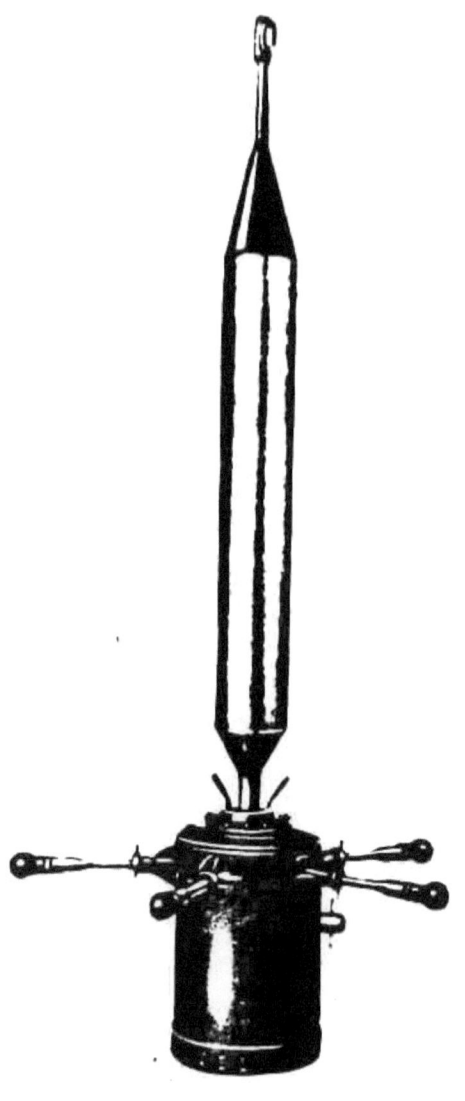

Figure 24 - EMS I Mine

Case material and thickness	Rolled steel - 1/8 in.
Method of laying	Surfaced U-boat or surface craft
Maximum laying height	10 ft.
Maximum laying speed	18 knots
Minimum water depths	10 ft. (approx.)
Arming time	15-20 minutes (soluble washer)
Self Destroying Mechanism	6-day clock

Figure 25 - EMS II Mine

The Characteristics of the three types of flotation gear are as follows:

EMS I employed a steel float designed to resemble a periscope, which is 66½ inches long and 6½ inches in diameter. When laid, the mine drifted with the upper portion of the float protruding several inches above the surface (figure 24).

EMS II employed a camouflaged, elliptical, steel float which was approximately 13 inches in diameter and eight inches high. This float was designed to ride flush with the surface of the water and to be invisible to pursuing vessels (figure 25).

EMS III employed a hemispherical, plexiglass float designed to resemble the dome of the Marder (midget submarine). The float was painted with a silhouette of the head and shoulders of a man as he would appear when operating a Marder. This float was approximately 25 inches in diameter and 14 inches high, and was designed to create the illusion of a partially submerged midget submarine with its dome exposed (figure 26).

Work on the EMS mine with periscope was started in 1941 and completed in 1942. It was designed to be laid by surfaced U-boats against all antisubmarine craft. These mines were stowed inside the submarine and had to be handed up through the conning tower when needed. They were assembled on deck and laid by hand by a two-man team. Because of handling and stowage difficulties, danger to laying personnel, and the extremely sensitive switch horns, only a few of these mines were used operationally, and the project was abandoned. During the same period, experiments were conducted with the EMT mine. This mine employed the same case as the EMS, but the periscope was replaced by a camouflaged elliptical float. Later, the EMT was designated the EMS II, and the EMS mine became EMS I. The EMT mine was primarily designed as a drifter for surface-craft laying; how-

CONTACT AND MOORED INFLUENCE MINES

Figure 26 - EMS III Mine

Figure 27 - FMC Mine Afloat

ever, it was abandoned at the same time as the EMS, for the same reasons.

In the fall of 1944 the German Sneak Attack Command (KDK) requested a floating mine with a plexiglass cupola attachment to resemble the dome of a midget submarine. These mines were intended to be laid by small motor boats or "Linsen" boats as decoys against patrol craft in heavily protected sea lanes. The first consignment of cupolas with the silhouette of a midget submarine operator for use with the EMS mine case was delivered in November 1944. Very few of these modified EMS mines were laid, because of operational difficulties and a small supply of cupolas. This modification had no German designation, and consequently, for purposes of identification, it is referred to as EMS III.

THE FM MINES

The German Fluss Minen (FM) mine series consisted of three types, FMA, FMB, and FMC. The FMB and FMC were completed and used operationally; the FMA was abandoned shortly after production was started.

The FM mines were small moored contact mines intended primarily for use in shallow waters of the Baltic and Black Seas and in rivers and estuaries. The FMA was designed and developed in 1920, but was abandoned in favor of FMB. The FMB was completed in 1926; however, none were produced during World War II, and only supplies on hand were laid. The FMC was developed and produced between 1926 and 1928, and only the supplies on hand at the beginning of the war were laid operationally. The FM series was considered ineffective, primarily because of the small explosive charge and the ease with which it could be swept. It was abandoned in favor of later improved models.

FMA Mine. The FMA was the first of the Fluss Minen series to be designed and developed as a moored, contact, surface-laid mine for use in the estuaries of the Baltic. It consisted of a hemispherical steel case approximately 22½ inches in diameter, with five chemical horns, a mooring buffer and wire cable mooring, and a charge of approximately 22 pounds of block-fitted hexanite. This mine was replaced by the FMB.

FMB Mine. The FMB was completed in approximately 1926. It was a surface-laid, moored, contact, chemical-horn mine using a cylindrical preset-type anchor (Anker mit absteckbarein ankertau).

FMC Mine. The FMC Mine was developed and completed between 1926 and 1928. It was a surface-laid, moored, contact, chemical-horn mine using the normal plummet-type anchor and containing a heavier charge than the FMB Mine.

Details. The FMB and FMC both used a wire cable mooring and a spring buffer; however, only the FMC took depth by plummet. Mooring tension pulled out the arming spindle, closed the mooring safety switch and the "A-E" switch, and tripped the booster-release lever to arm the mines. The "A-E" switch in these mines served only to open or close a switch in the horn circuit. The FMB had two lifting eyes welded to the upper hemisphere. FMC had one lifting eye welded to the upper hemisphere and three anchor-securing lugs; one on the upper hemisphere and two on the lower hemisphere. There are two designations for the FMB mine: FMB and FMB(35). These differ only in the weight of explosive, 28 pounds and 44 pounds, respectively. The FMB and FMC differ from each other as follows:

GERMAN UNDERWATER ORDNANCE—MINES

Figure 28 - FMB Mine

Figure 29 - FMC Mine

FMB Mine

    Diameter of mine case   26 in.

    Weight of charge   28 lb. or 48 lb.

    Number of horns   5: one, in center of upper hemisphere; 4 equally spaced around upper hemisphere

    Depth setting   preset

    Minimum depth setting   ----

    Mooring cable   25 ft. long 1/2-in. diameter
                              50 ft. long 1/2-in. diameter

FMC Mine

    Diameter of mine case   30 in.

    Weight of charge   88 lb.

    Number of horns   same as FMB

    Depth setting   1 ft. to 15 ft.

    Minimum depth setting   13 ft. plus length of depth-setting mechanism

    Mooring cable   160 ft. long - 7/16-in. diameter
                       475 ft. long - 5/16-in. diameter

The only self-disarming device is the mooring safety switch, which is designed to disarm the mine by opening the firing circuit upon release of mooring tension.

THE OMA MINES

The OMA (Oberflachen Mine A) was a moored, contact, surface mine the development of which was undertaken in late 1942. It utilized a novel type of mine case and was designed in five models, which were designated OMA I, OMA II, OMA III, OMA IV, and OMA/K. Of these five models only the OMA I and OMA/K were used operationally. The OMA II and OMA III were abandoned in the preliminary development stage, and the OMA IV was unperfected at the close of the war.

CONTACT AND MOORED INFLUENCE MINES

Figure 30 - OMA/K Mine

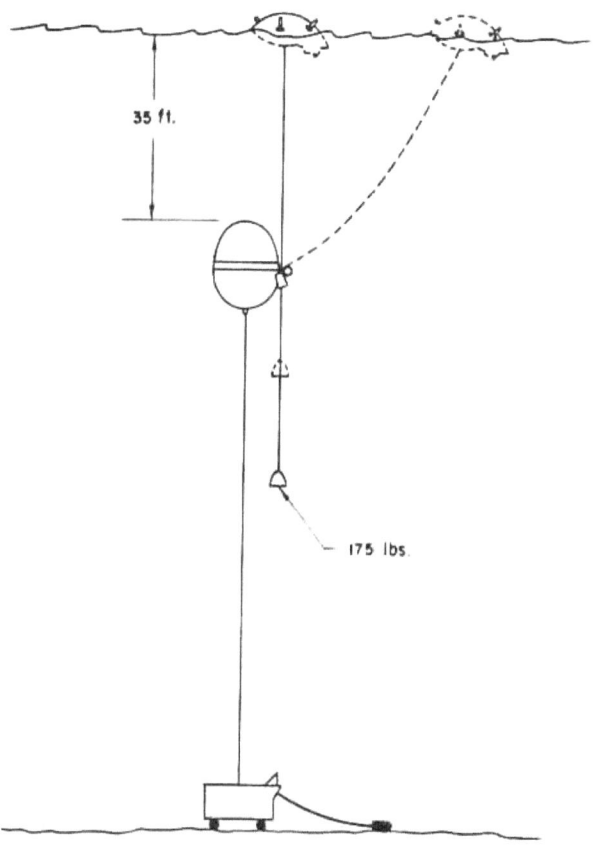

Figure 31 - OMA IV Mine

The first moored, contact, surface mine developed by the Germans was a jury-rig affair designated OMA/K; it was laid in 1942. In the fall of the same year the development of a more efficient mine was undertaken, this type being designated OMA I. Original requirements for the OMA I called for its use in water depths up to 85 feet, but these requirements were subsequently increased to 325 feet. The development of this mine was completed in February 1943, and it was laid in the fall of 1943.

In 1944 the German Navy Department, alarmed by the large number of OMA I mines breaking loose, reduced the water-depth requirements to a more suitable depth of 50 feet and requested an appropriate modification of the mine. The resulting modification was designated OMA/K.

With the development of OMA/K the problem of preventing excessive drifters was successfully met for depths up to 50 feet. (The OMA/K, when tested in the extremely rough waters of the Elbe Estuary, lasted for more than five months.) However, since the buoyancy of the OMA case precluded the use of a double chain mooring in depths over 50 feet, the problem remained unsolved for use in greater depths.

Late in 1943 the development of a moored, contact, surface mine for depths up to 985 feet was undertaken. This mine was to consist of an EMC anchor, a large steel float with a guide arrangement, an OMA Mine case with a normal spindle-type arming and disarming switch, and a 175-pound weight. A schematic representation of the assembly is shown in figure 31. This assembly was designated OMA IV. Because of the relative complexity and low priority of this mine, its development was incomplete at the close of the war.

Neither the OMA I nor the OMA/K was fitted with the normal disarming devices found in moored mines. This was due to the fact that these mines had a slack mooring cable and could not be fitted with switches of the normal mooring-spindle or hydrostatic types. Therefore, since these mines broke adrift fairly often, their use created an unacceptable hazard to German shipping and shore installations. To remedy this situation the development of the OMA II and OMA III was undertaken. The OMA II was to incorporate a mechanical-type disarming device and the OMA III an electrical type. However, both types presented a great number of difficulties and were abandoned in the preliminary development stage.

Since disarming devices of the normal type could not be applied to the OMA I and OMA/K, the German Navy Department ordered that they be fitted with a ZE IVa. (60-day disarming clock).

The five models of this mine utilized the same type of mine case. The characteristics of the case are as follows:

RESTRICTED

GERMAN UNDERWATER ORDNANCE—MINES

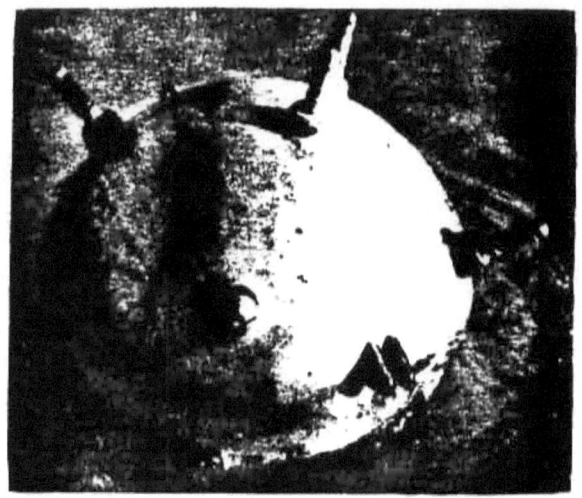

Figure 72 - UMA Mine Afloat

Method of firing	chemical horns
Total weight	352 lb.
Weight of charge (cast)	66 lb.
Positive buoyancy	440 - 22 lb.
Total height	31½ in.
Diameter less skirt	42½ in.
Delay in arming	20-35 minutes (Soluble washer)

The significant design feature of this mine case was the steel skirt fitted to the underside of the case. The purpose of this skirt was to eliminate dipping of the mine.

The OMA I and the OMA/K were identical except for the types of mooring and anchors used. The OMA I mooring consisted of a length of chain secured on one end to the mine case and on the other end to a length of steel cable leading from the mine anchor. The OMA/K mooring consisted of a 35-foot bight of chain secured on one end to the mine case and on the other end to a length of chain leading from the mine anchor. The anchors were substantially the same, except that the cylindrical mine stool around which the mooring was coiled was higher on the OMA I than on the OMA/K.

THE UM MINES

The U-Bootsabwehrmine (UM) mine series consisted of three types of moored, contact, surface-laid mines, the UMA, UMA/K and UMB.

The UMA Mines. The first of this series was developed in 1928. It was a moored, contact, chemical- and switch-horn mine, laid by surface craft, intended for use primarily against submarines. In 1936 the prototype

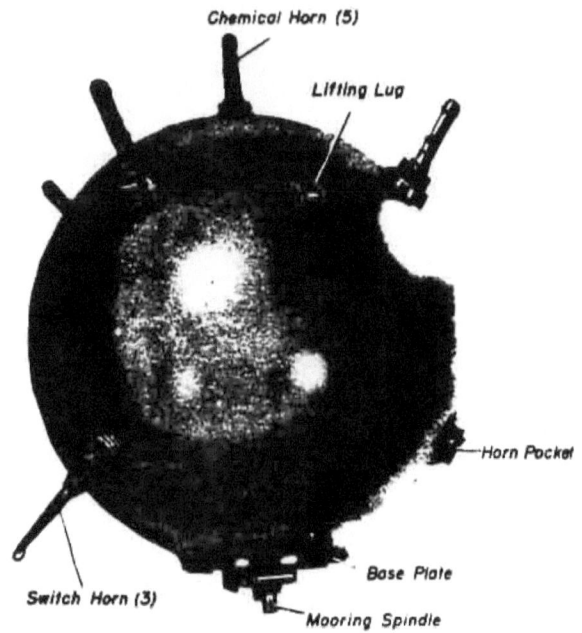

Figure 73 - UMA Mine

of this series became the UMA I when, after slight modifications in the case and mooring, the UMA II was introduced. In 1939 the UMA III appeared. Basically, it was the same as its predecessors, with improvements in the anchor and mooring spindle.

Description of Case

Shape	Spherical
Material	Steel
Diameter	32 in.
Charge	66 lb. block-fitted hexanite

Description of External Fittings

Horns	Eight: one chemical, in center of upper hemisphere; four, chemical, equally spaced around upper hemisphere, 15½ in. from center; three, switch, equally spaced around lower hemisphere, 17 in. from center
Base plate	Standard type UMA
Lifting eye	One 19 in. from center of upper hemisphere

CONTACT AND MOORED INFLUENCE MINES

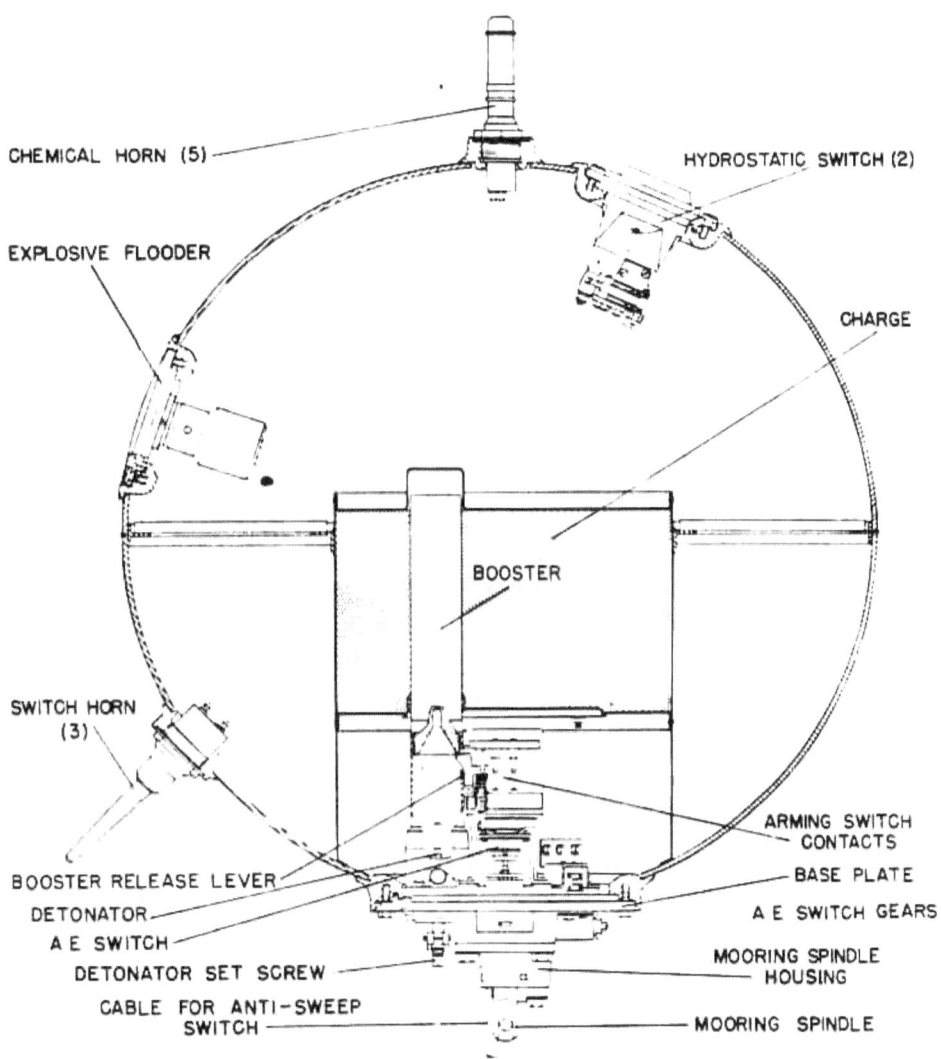

Figure 34 - UMA Mine - Cross Section

Lifting lug   One, 180° from lifting eye, 19 in. from center of upper hemisphere

Operation. The depth of the mine case is preset and determined by the amount of mooring cable used. Mooring tension pulls out the mooring spindle, closing the mooring safety switch, tripping the booster-release lever, and the mine is armed.

The only self-disarming device is the mooring safety switch which is designed to disarm the mine by opening the firing circuit upon release of mooring tension.

The German UMA/K mine assembly was designed and developed in late 1942 to defend the sea approaches to Northern Europe from Allied attack. It replaced the EMG mine assembly, which was considered unsatisfactory for the defense of such waters, and was used operationally until 1944, when it was in turn replaced by the CMA/K.

The UMA/K consists of a normal UMA mine case with the lower horns blanked off, a cast-iron 110-pound weight and a cast-iron anchor weighing 500 pounds. The assembly of the above components before and after laying is shown in figures 35 and 36.

The assembly is so designed that the mine case remains on the surface, regardless of the stage of the tide. It may be laid in depths of 85 to 165 feet. For depths of 85 to 150 feet, a 65-foot mooring cable is used between the mine case and weight; for depths of 100 to 165 feet, the mooring-cable

RESTRICTED

GERMAN UNDERWATER ORDNANCE—MINES

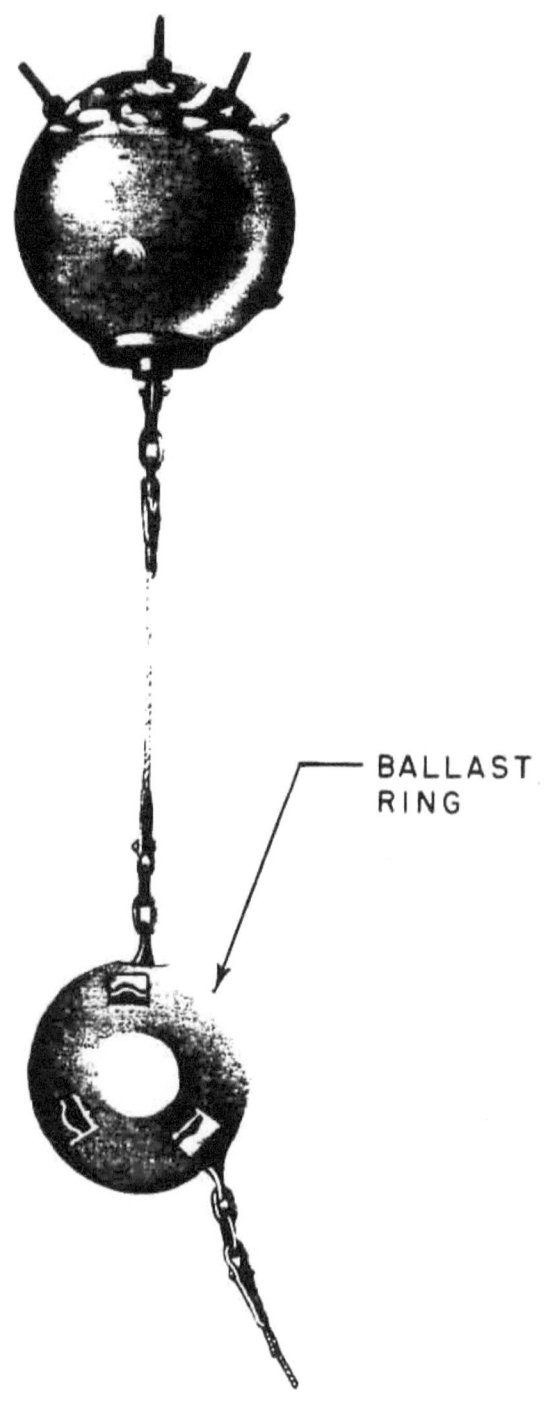

Figure 35 - UMA/K Mine

Figure 36 - UMA/K Mine

length is increased to 85 feet. In both cases, the mooring cable between the anchor and weight is 165 feet in length.

The UMB Mine. The UMB mine was first started in 1941, and it appeared in four forms. It was a moored, contact, chemical- and switch-horn mine, laid by surface craft.

It was an offensive or defensive mine, for use in maximum water depth of 500 feet, against surface craft and submarines. The maximum depth of case when moored is 110 feet.

The Four Forms

UMB with tombac tubing; 1941

UMB with 5-foot chain and mechanical cutter; 1943

UMB with 5-foot chain, mechanical cutter, and cork-floated snag line; 1943/44

UMB with 5-foot chain and two improved mechanical cutters; 1944

Description of Case

Shape	Spherical
Material	Steel
Diameter	33.5 in.
Charge	90 lb. block-fitted hexanite

CONTACT AND MOORED INFLUENCE MINES

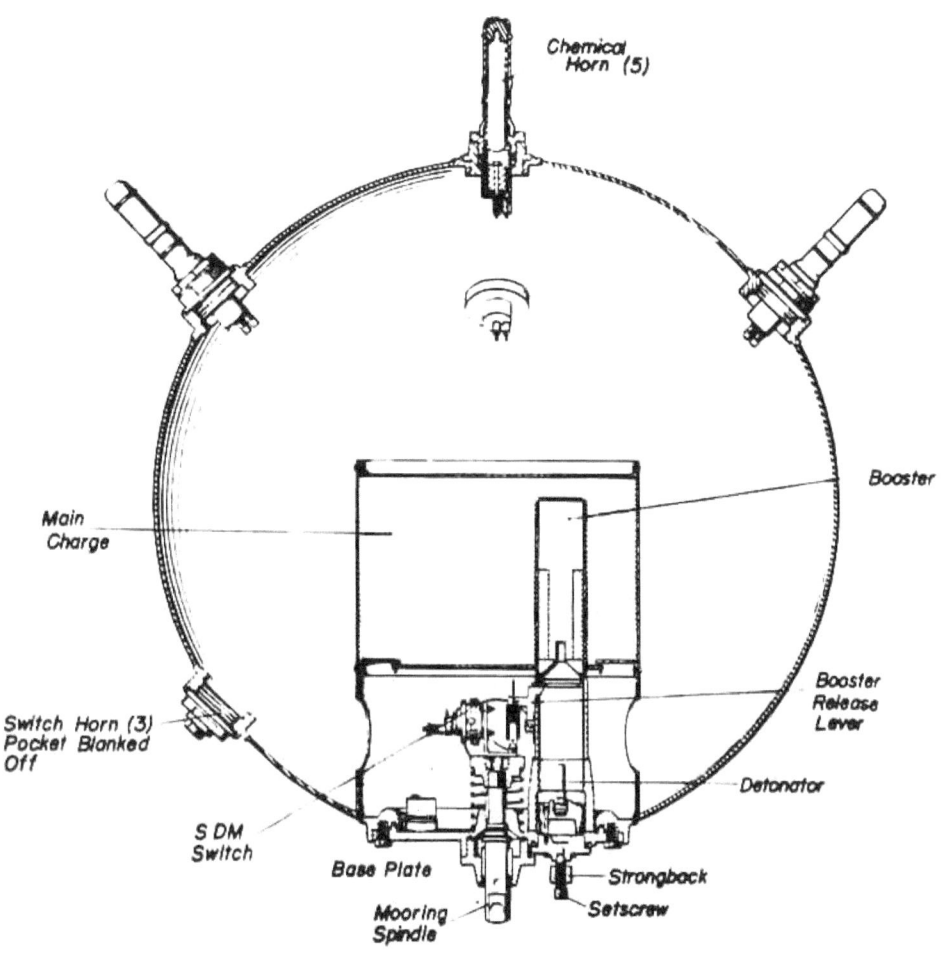

Figure 37 - UMF Mine - Cross Section

Description of External Fittings

Horns	Eight; one, chemical, in center of upper hemisphere; four, chemical, equally spaced around upper hemisphere, 17 in. from center; three, switch, equally spaced around lower hemisphere, 17 in. from center
Base plate	Standard type UMB
Hydrostatic switch covers	Two: 6.5-in. diameter; one, 7.5-in. from center of upper hemisphere; one, 17 in. from center of lower hemisphere
Explosive flooder cover	6.5 in diameter, 23 in. from center of upper hemisphere
Securing lugs	Three; one, 20 in. from center of upper hemisphere; two 20° apart 12 in. from center of lower hemisphere
Snag line (Optional)	79 feet long, secured to center of three-foot length of wire connecting two switch horns. When the mine is so rigged, the chemical horn directly above is blanked off.

RESTRICTED

GERMAN UNDERWATER ORDNANCE—MINES

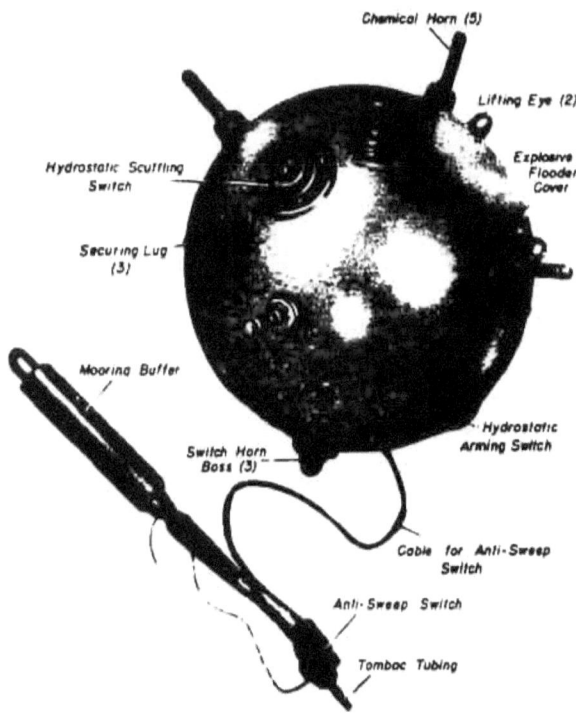

Figure 38 - UMB Mine

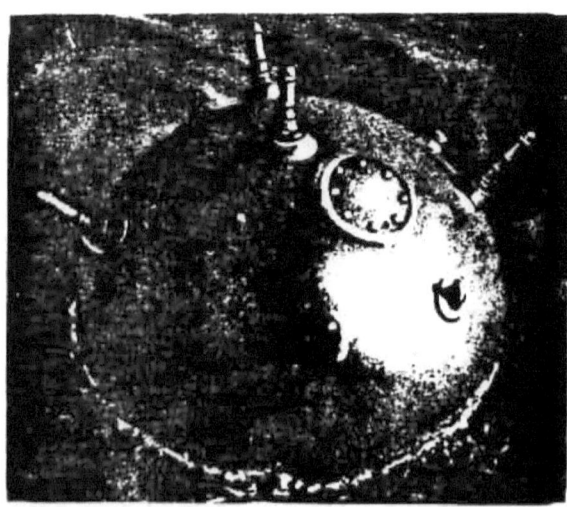

Figure 39 - UMB Mine Afloat

The hydrostatic scuttling switch on the upper hemisphere is an anti-shallow-plant hydrostat which controls a double-pole switch, normally made to one of its contacts. The hydrostat may be set to any one of four depths: 0, 5, 10, or 15 meters. If, upon laying, the mine moors at a depth shallower than that set on the hydrostat, the ex-

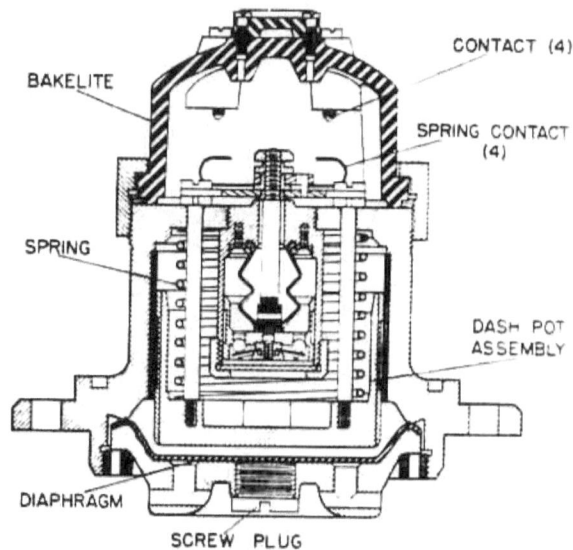

Figure 40 - Hydrostatic Arming Switch for UMB Mine

plosive flooder will fire upon closure of the mooring safety switch. If the mine moors correctly, (i.e., at a depth greater than that set on the hydrostat) the switch changes over to the other contact, permanently breaking the flooder circuit.

The hydrostatic arming switch, figure 40, on the lower hemisphere is designed to open or close the firing circuit when the mine rises above or descends below a depth of six feet. A glycerine-filled dash-pot delays the action of the switch for a period of 20 seconds. A screw plug, fitted to the center of the switch cover, is painted white when the switch is rigged to operate as described above. If the plug is painted red, however, it indicates that the switch has been closed during assembly, being held in that position by a special extension arm added to the screw plug. In this case, the switch will not open under any circumstances.

Operation. The mine takes depth by plummet. The hydrostatic switch closes in six feet of water (if red screw plug is fitted, switch is permanently closed) and, if the mine moors at a depth greater than that set on the anti-shallow-plant hydrostatic switch, the flooder circuit is broken. Dissolution of a soluble plug allows mooring tension to pull out the mooring spindle, closing the mooring safety switch and tripping the booster-release lever, and the mine is armed. A spring-loaded detent is usually fitted to lock the mooring spindle out.

The mine has standard chemical or switch-horn firing. An additional firing method may be incorporated by fitting a "tombac" anti-sweep tubing to the mooring cable. Upward movement of this tubing along the mooring cable, such as might be caused by a sweep wire contacting it, will close a switch on the

CONTACT AND MOORED INFLUENCE MINES

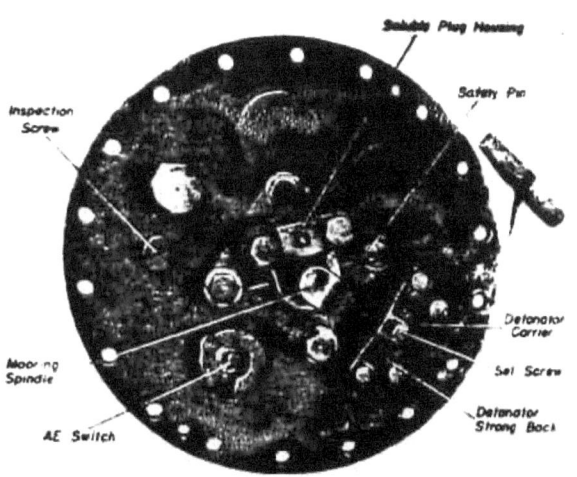

Figure 41 - Base Plate Type EMC II (Exterior)

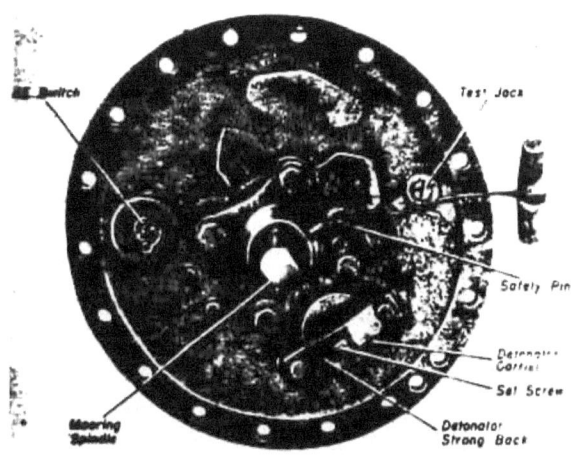

Figure 42 - Base Plate Type UMA (Exterior)

tubing and fire the main charge. Mines fitted with snag lines will not normally be fitted with the "tombac" anti-sweep device nor the locking detent on the mooring spindle.

The mooring safety switch is designed to disarm the mine by opening the firing circuit upon release of mooring tension except when the detent is fitted. The hydrostatic arming switch is also designed to break the firing circuit if fitted with a white screw plug.

BASE PLATES

All German moored contact mines are spherical or have cases consisting of two hemispheres joined by a cylindrical mid-section. The cases are of mild steel, vary in diameter from 26 inches to 46 inches, and are loaded either with cast or block-fitted Hexanite. Chemical and switch horns are employed, either singly or in combination.

Mines of this type usually depend on mooring tension for arming and disarming, these processes being controlled through the mooring spindle on the base plate. General characteristics are given below:

All base plates are fitted with straight-shank mooring spindles which are withdrawn by mooring tension against tension of a coil spring mounted on the inside of the base plate.

Withdrawal of the mooring spindle performs the following functions:

It trips the booster release lever.

It arms the self destroying mechanism.

It closes the mooring safety switch.

The booster release lever is mounted in the booster tube and is connected, by means of a mechanical linkage to the mooring

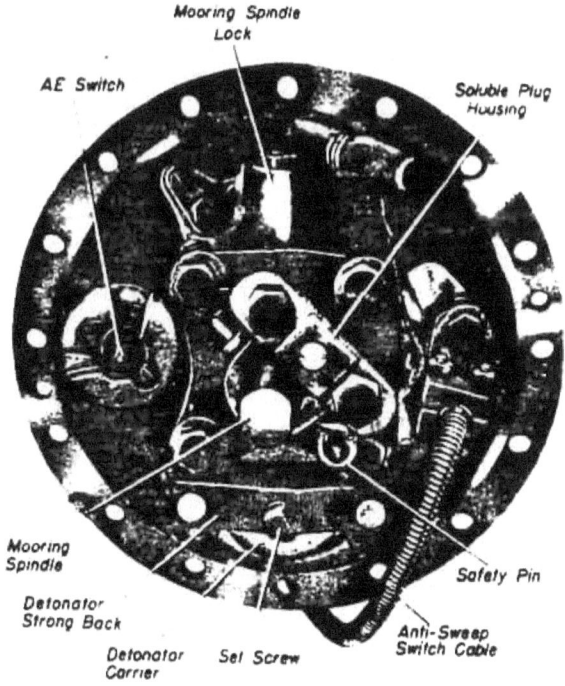

Figure 43 - Base Plate Type UMB (Exterior)

spindle. The lever holds the booster in the "Safe" position above the detonator until the mooring spindle is withdrawn, at which time the lever is tripped and the booster is freed to drop over the detonator.

The self-destroying mechanism may be either an electrochemical internal horn (often referred to as the "eighth horn") used in the Type EMC II base plate, or a

RESTRICTED

GERMAN UNDERWATER ORDNANCE—MINES

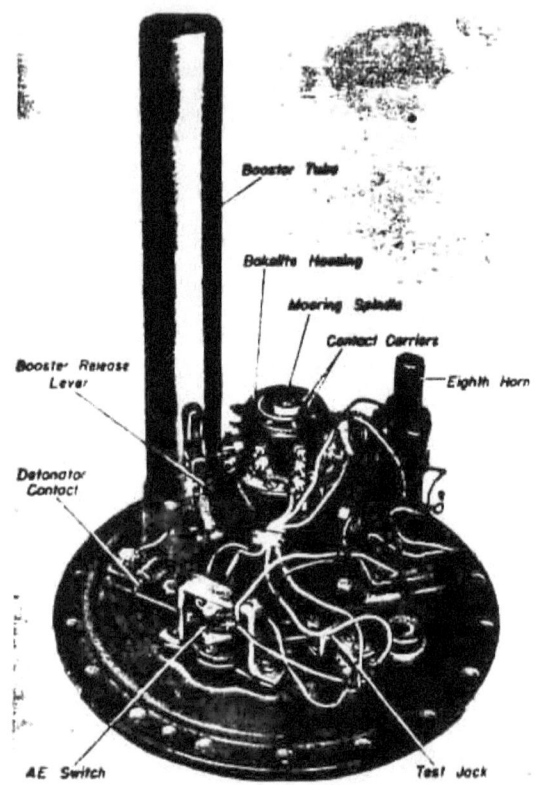

Figure 44 - Base Plate Type EMC II (Interior)

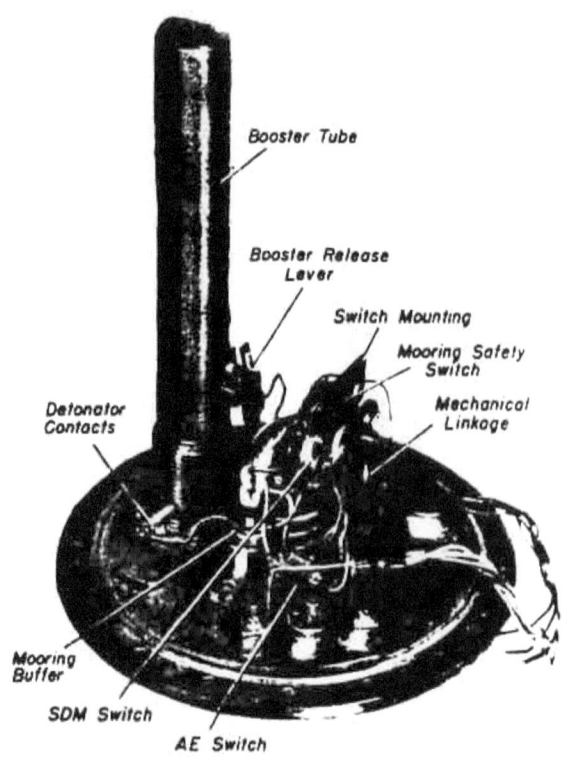

Figure 45 - Base Plate Type UMA (Interior)

rotary, two-position switch used in the Type EMC I or UMA base plates.

The horn-type self-destroying mechanism is mounted in a casting secured to the inside of the base plate by four bolts. Its operation is controlled by a mechanical linkage connected to the mooring spindle. Withdrawal of the spindle allows a cocking pin to move the armed position, and retraction of the spindle pivots the cocking pin and releases a spring-loaded firing pin, which shatters the electrolyte ampoule. The electrolyte then runs into a battery, energizing it, producing a momentary current sufficient to fire the detonator and main charge, if the self destroying mechanism is in the firing circuit.

The switch-type self-destroying mechanism is mounted on a bracket on the inner end of the mooring spindle and is connected to the base plate by a mechanical linkage. Withdrawal of the spindle carries a small pin into position behind a cam. Retraction of the spindle carries the cam back with the pin and closes the switch.

The various base plates use the following types of mooring safety switches:

With Base Plate Type EMC - a switch consisting of four contacts, two of which are mounted on the mooring spindle and two on the mooring spindle housing. Withdrawal of

the spindle makes the contacts, arming the horn circuit. Retraction of the spindle breaks the contacts, disarming the horn circuit.

With Base Plates Types EMC I and UMA - a two-position rotary switch mounted on a bracket on the mooring spindle and connected to the base plate by a mechanical linkage. Withdrawal of the mooring spindle closes the switch, arming the horn circuit. Retraction of the spindle opens the switch, disarming the horn circuit.

With Base Plate Type EMC II - a switch consisting of two main parts: a cylindrical, bakelite housing mounted on the base plate and enclosing the inner end of the mooring spindle; two bakelite-covered brass cylinders mounted one above the other on the inner end of the mooring spindle. The latter are fitted with brass contact pieces, and the former with spring-loaded contacts. Withdrawal of the spindle pulls down the two cylinders with respect to the housing, so that the contact pieces make their respective contacts, arming the horn and self-destroying mechanism circuits. Retraction of the spindle breaks the upper set of contacts, disarming the horn circuit. The lower set is in the SDM circuit and remains closed, being locked by a spring-loaded detent.

With Base Plate Type UMB - a switch consisting of eight contacts, four of which are

CONTACT AND MOORED INFLUENCE MINES

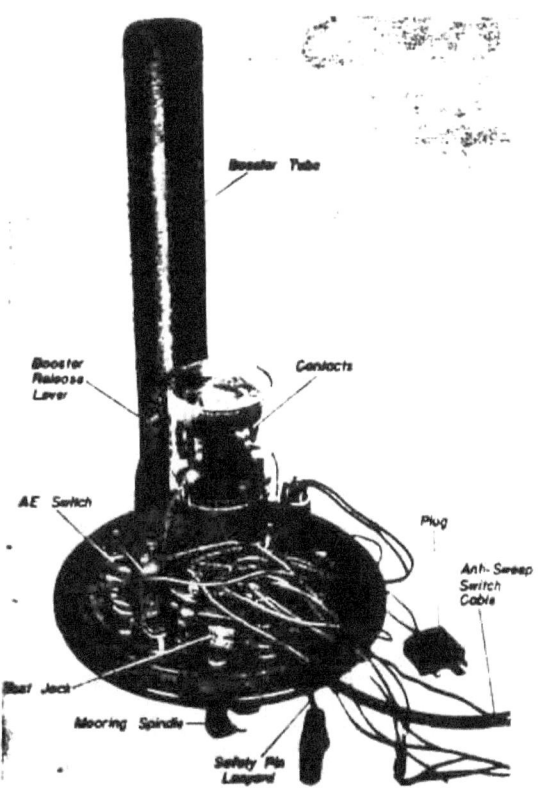

Figure 46 - Base Plate Type UMB (Interior)

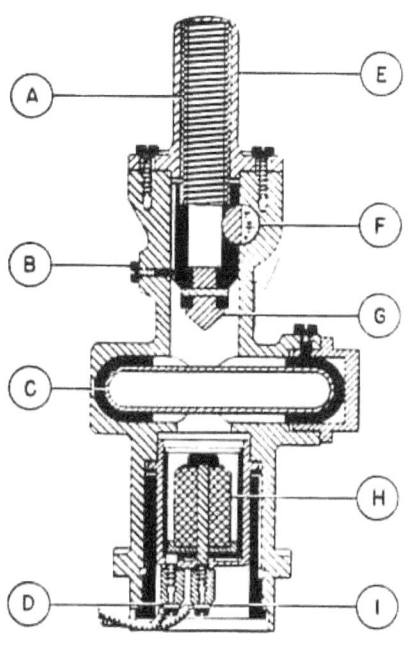

A - STRIKER SPRING
B - STRIKER GUIDE PIN
C - GLASS VIAL
D - NEGATIVE CONTACT
E - STRIKER SPRING HOUSING
F - STRIKER RELEASE PAWL
G - STRIKER
H - HORN BATTERY
I - POSITIVE CONTACT

Figure 47 - Eighth Horn

mounted on a cross-head on the mooring spindle and four on the mooring-spindle housing. Withdrawal of the spindle makes the contacts, arming the horn circuit. Retraction of the spindle breaks the contacts, disarming the horn circuit. However, the mooring spindle is designed to lock in the "out" position.

A detonator carrier is fitted in a well located externally on the base plate beside the mooring spindle, and is held in place by a strongback and a single setscrew. The screw fits into a boss on the detonator carrier and is secured by a U-pin which fits into an annular groove on the setscrew. Two spring-loaded contacts are mounted on the inside of the base plate, extending vertically upward and then bending at an angle of 90° to enter the booster tube. These contacts make similar contacts on the detonator carrier when it is inserted in the booster tube.

A spindle which controls an internal, two-position rotary switch is mounted at either 90° or 135° from the detonator carrier. A red arrow is stamped on its face to indicate the switch setting and the letters A and E are stamped on the part of the base plate adjoining. This switch is in the circuit of the SBM, except in base plate Type UMB, where it is in the circuit of the UMB, where it is in the circuit of the "tombac" anti-sweep device. If the arrow points to A (painted white), the switch is open and the self-destroying mechanism or "tombac" is not in the circuit. If the arrow points to E (painted red), the SDM or "tombac" is in the circuit and both should operate as designed.

A soluble plug holder may be found alongside the mooring spindle, secured by a strongback. A black, plastic disc, about a half inch in diameter, is fitted in the strongback. Withdrawal of the mooring spindle upon dissolution of the soluble plug pushes this disc out of the strongback. Note that the presence or absence of this disc provides a positive means of determining whether or not the mine has ever armed.

In some cases, the following additional base plate fittings may be found:

A gland for connecting a lower antenna or "tombac" anti-sweep device

A slotted screw plug for applying a circuit tester

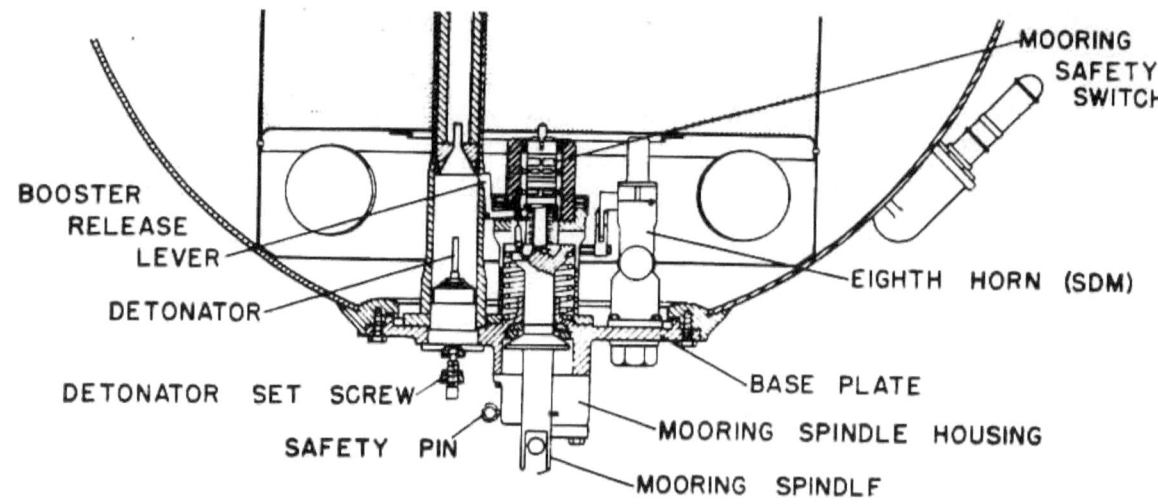

Figure 48a - Diagram of Base Plate Type EMC II

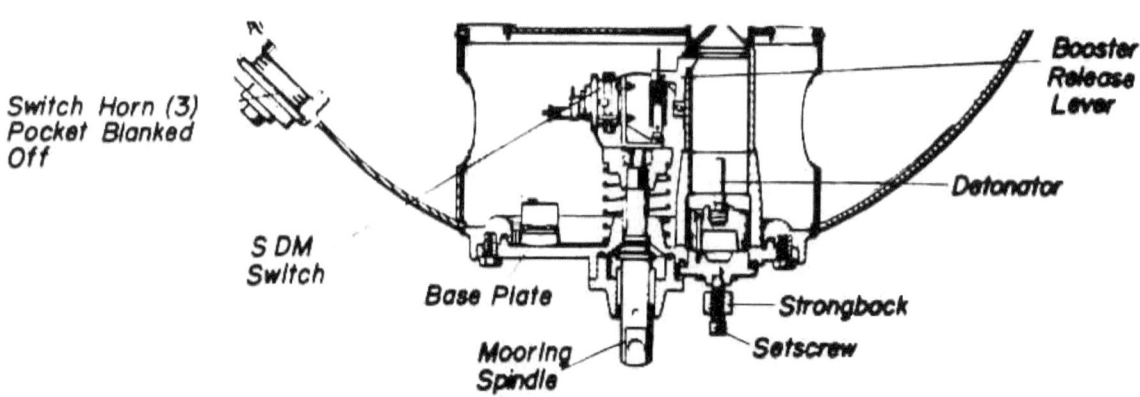

Figure 48b - Diagram of Base Plate Type UMA

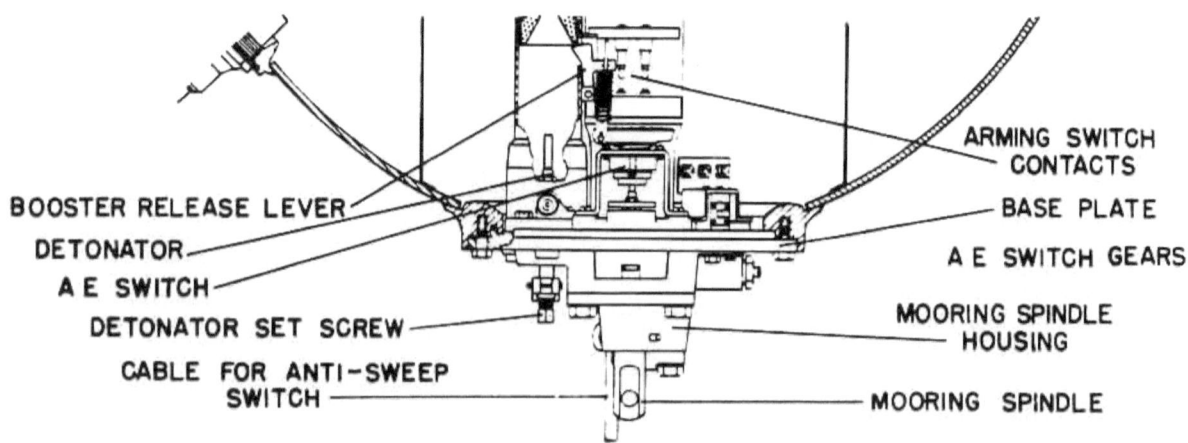

Figure 48c - Diagram of Base Plate Type UMB

CONTACT AND MOORED INFLUENCE MINES

TABLE OF BASE PLATES

Base Plate	Where Used	Diameter (inches)	Material	Mooring Spindle Delay	Type of Booster Tube	SDM (self-destroying mechanism)	How Secured	Remarks
Type EMC	EMC EMD	15	Gun metal	Two oil dash pots	8 1/2 in. long, open at top	Chemical horn mounted in top of mooring spindle tube	Secured by 20 bolts	Considered obsolete
Type EMC I	EMC EMD	15	Steel	Two oil dash pots	18 in. long, closed at top	Rotary two-position switch	Secured by 19 bolts	Considered obsolescent
Type C EMC II	EMC EMD	15	Steel	Soluble plug	18 in. long, closed at top	Chemical horn mounted beside mooring spindle	Secured by 20 bolts	Fitted with lower antenna gland 180° from booster tube; gland blanked off with a hexagonal cap, if not, antenna is fitted
Type UMA	UMA	11 1/2	Steel	Two oil dash pots	15 in. long, closed at top	Rotary two-position switch	Secured by 19 bolts	Considered obsolete
Type UMB	UMB	11 1/2	Steel	Soluble plug	15 in. long, closed at top	None	Secured by 16 bolts	Fitted with Tombac firing device gland, 180° from A-E switch and with mooring spindle locking detent, 180° from booster tube

Figure 48d - Base Plates - Table of Data.

RESTRICTED

GERMAN UNDERWATER ORDNANCE—MINES

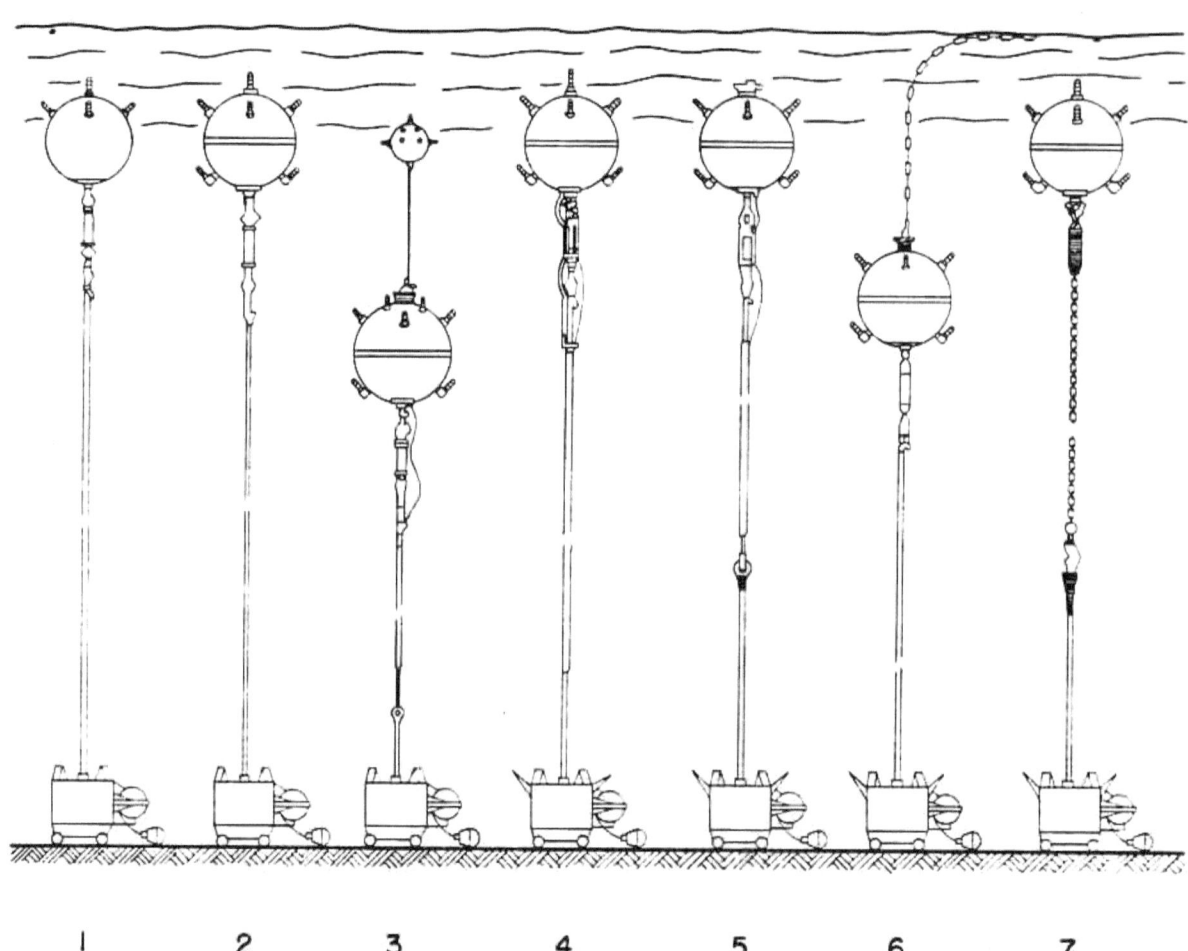

KEY TO FIGURE 49a

1. EMD I, II
2. EMC I
3. EMC II - Upper and lower antenna
4. EMC II - Tombac tubing
5. EMC II - lower antenna
6. EMC II - Cork-floated upper antenna
7. EMC II - Chain mooring

Figure 49a - Operational Mines and Sweep Obstructors

CONTACT AND MOORED INFLUENCE MINES

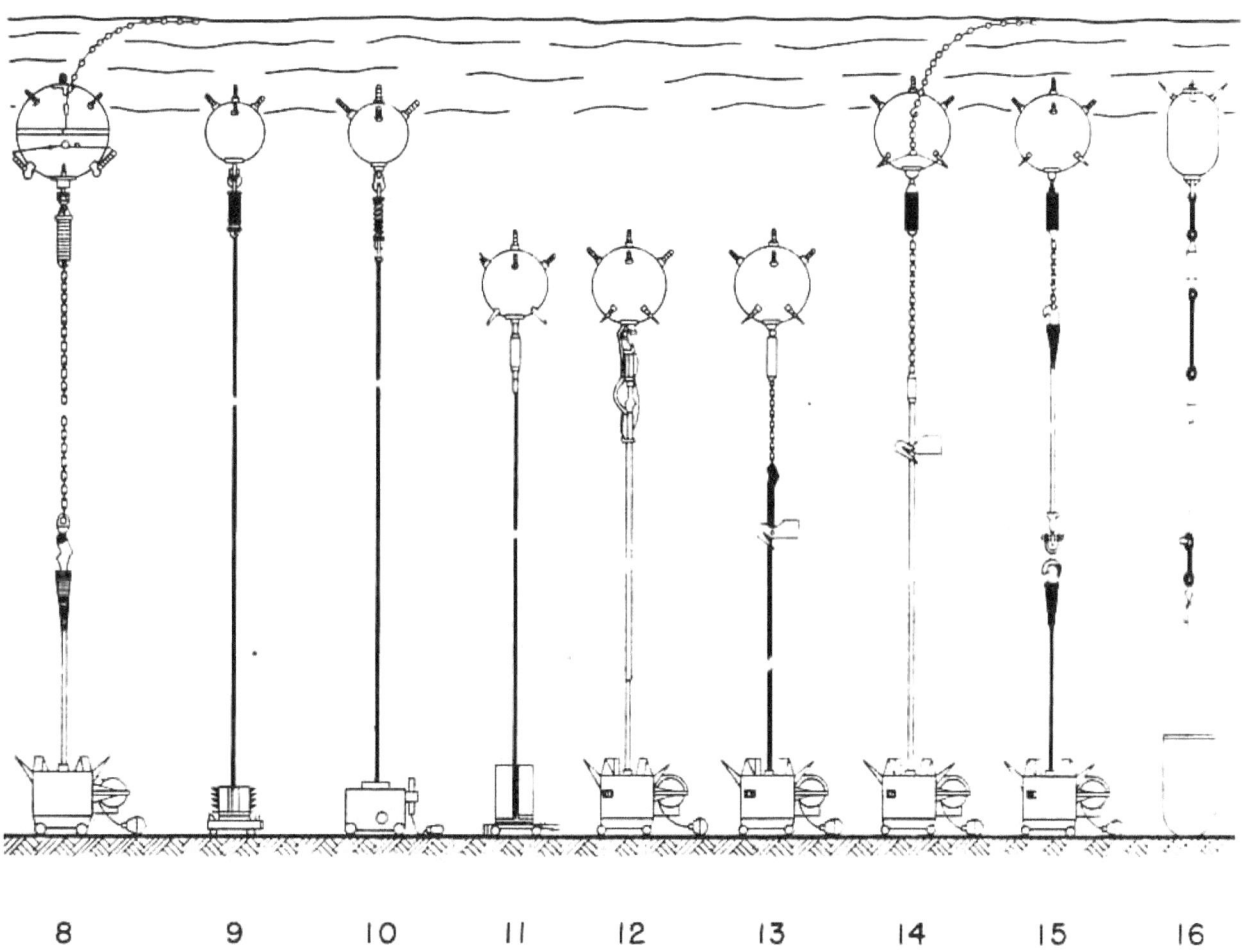

KEY TO FIGURE 49b

8. EMC II - Snag line and chain mooring
9. FMB
10. FMC
11. UMA
12. UMB - Tombac tubing
13. UMB - Chain mooring and mechanical cutter
14. UMB - Snag line, chain mooring, and mechanical cutter
15. UMB - Chain mooring and two cutters
16. BMC

Figure 49b - Operational Mines and Sweep Obstructors (Continued)

GERMAN UNDERWATER ORDNANCE—MINES

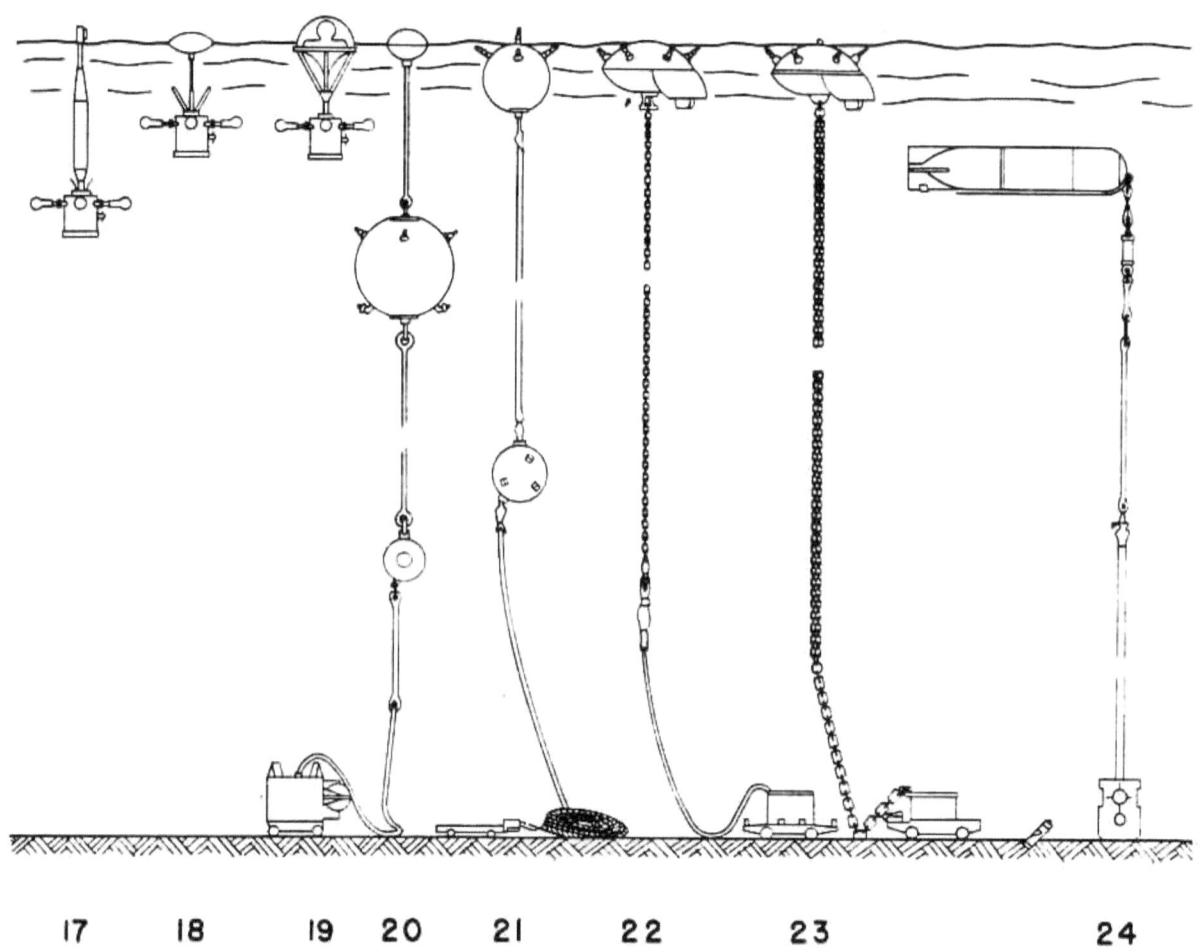

KEY TO FIGURE 49c

17. EMS I - Periscope type
18. EMS II - Float type
19. EMS III - Cupola type
20. EMG
21. UMA/K
22. OMA I - Single chain
23. OMA/K - Double chain
24. EMF

Figure 49c - Operational Mines and Sweep Obstructors (Continued)

CONTACT AND MOORED INFLUENCE MINES

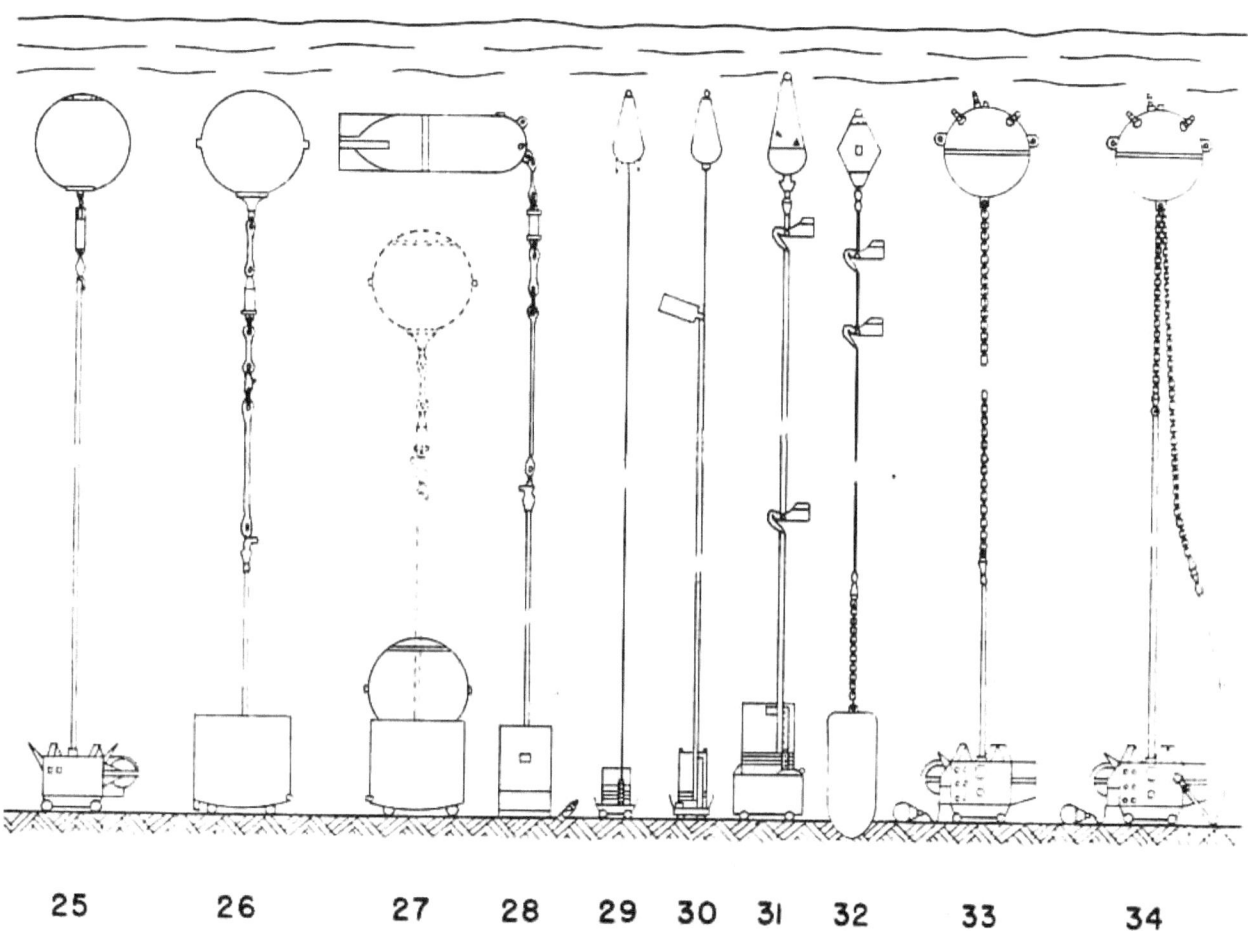

KEY TO FIGURE 49d

25. TMA
26. SMA
27. SMC
28. LMF
29. Explosive anti-sweep float C
30. Explosive anti-sweep float D
31. Mechanical anti-sweep float
32. BRB
33. EMR
34. EMR/1

Figure 49d - Operational Mines and Sweep Obstructors (Concluded)

GERMAN UNDERWATER ORDNANCE—MINES OP 1673A

Designation	Stage of Development	Type	Method of Laying	Method of Firing	Wt. of Charge (pounds)	Case Material	Dimensions Diam.	Length
BMC	Operational	Moored	Aircraft	Le-Clanche horn	120	Steel	26	44
BMC/S	Operational	Moored	Surface	Le-Clanche horn	120	Steel	26	44
EMA	Operational	Moored	Surface & submarine	Chemical horn	330	Steel	34	46
EMB	Operational	Moored	Surface & submarine	Chemical horn	480	Steel	34	46
EMC I	Operational	Moored	Surface	Chemical horn	660	Steel	46	48
EMC II	Operational	Moored	Surface	Chemical horn lower and upper antenna	660	Steel	46	48
EMC II (Tombac Tubing)	Operational	Moored	Surface	Chemical horn and tombac tubing	550 to 660	Steel	46	48
EMC II (Lower Antenna)	Operational	Moored	Surface	Chemical horn and lower antenna	550 to 660	Steel	46	48
EMC II (Cork-Floated Upper Antenna)	Operational	Moored	Surface	Chemical horn and antenna	630 to 660	Steel	46	48
EMC II (Chain Mooring)	Operational	Moored	Surface	Chemical horn	550	Steel	46	48
EMC II (Snag-Line and Chain Mooring)	Operational	Moored	Surface	Chemical horn	550	Steel	46	48
EMD I/II	Operational	Moored	Surface	Chemical horn	330	Steel	40	40
EMG	Operational	Moored	Surface	Chemical horn	660	Steel	46	48
EMK	Development	Moored	Surface	Chemical horn and/or influence unit	660	Steel	-5	50

Figure 50 - Contact Mines - Table of Data (continued on next page)

CONTACT AND MOORED INFLUENCE MINES

Total Wt. (with anchor)	Remarks
1430	Developed by SVK for use by the Luftwaffe. Is only SVK mine in which booster and detonator are married prior to laying.
1430	Same as BMC/S but for laying by E-boat.
2000	A World War I mine which existed in two models, one for laying by surface craft and the other by special vertical-shaft U-boats. Stocks of the surface-laid model which remained on hand in 1939 were laid during World War II.
2200	Same as EMA except for weight of charge.
2375	This was the standard contact mine, and it was in very wide use in the various forms listed.
2400	Differs from Model I by use of upper and lower antenna and improvements to base plate.
2500	Tombac tubing added to mooring cable to provide protection against submarines and to act as anti-sweep device.
2400	This was first type of EMC mine to which ZE III (80-day clock) scuttling clock was fitted.
2400	Developed to defend against shallow-draft vessels such as PT-boats.
2500	Utilizes 18-foot length of 16-mm chain as anti-sweep device.
2500	Developed to defend against shallow-draft vessels.
2100	Developed at same time as EMC but intended for use against surface craft only. Consequently it had no lower horns. Is similar in all other respects to EMC except for size.
2680	A constant-depth assembly designed to protect against shallow-draft vessels. Uses EMC case, float, and special ballast weight.
Unknown	This mine was first attempt to develop a moored mine radically different from the EMC.

GERMAN UNDERWATER ORDNANCE—MINES

Designation	Stage of Development	Type	Method of Laying	Method of Firing	Wt. of Charge (pounds)	Case Material	Dimensions Diam.	Dimensions Length
EMS I	Operational	Drifting	Surface or submarine	Switch horn	30	Steel	15	75
EMS II	Operational	Drifting	Surface or submarine	Switch horn	30	Steel	15	30
EMS III	Operational	Drifting	Surface or submarine	Switch horn	30	Steel	15	40
EMU	Development	Moored	Surface	Chemical horn and/or influence unit	220	Steel	40	45
FMA	Abandoned	Moored	Surface	Chemical horn	22	Steel	23	25
FMB	Operational	Moored	Surface	Chemical horn	30 or 44	Steel	26	29
FMC	Operational	Moored	Surface	Chemical horn	88	Steel	30	33
KMA	Operational	Ground	Surface	Chemical horn	165	Concrete	47	47
OMA I	Operational	Surface	Surface	Chemical horn	66	Steel	42	41
OMA/K	Operational	Surface	Surface	Chemical horn	66	Steel	42	41
OMA II	Abandoned	--	--	--	--	--	--	--
OMA III	Abandoned	--	--	--	--	--	--	--
OMA IV	Development	--	--	--	--	--	--	--
UMA	Operational	Moored	Surface	Chemical horn & switch horn	66	Steel	29	29
UMB (Tombac Tubing)	Operational	Moored	Surface	Chemical horn & switch horn	88	Steel	33	33
UMB (Chain & Cutter)	Operational	Moored	Surface	Chemical horn & switch horn	88	Steel	33	33
UMB (Snag-Line, Chain & Cutter)	Operational	Moored	Surface	Chemical horn & switch horn	88	Steel	33	33
UMB (Chain & Two Cutters)	Operational	Moored	Surface	Chemical horn & switch horn	88	Steel	33	33

Mines - Table of Data (concluded)

Total Wt. (with anchor)	Remarks
100	Utilizes a flotation chamber made to resemble a periscope.
65	Same as Model I but a camouflaged float used in place of "periscope".
80	Same as Model I but plexi-glass float used in place of "periscope".
Unknown	Smaller model of EMI.
Unknown	Developed about 1920 for use in Baltic; abandoned in favor of FMB.
600	Primarily intended for use in shallow waters of Baltic. Uses preset-type anchor.
920	Similar to FMC except for size. Utilizes normal plummet-type anchor.
2200	Intended to defend against landing barges.
2380	Only mine specially designed for watching on surface.
2380	Same as OMA I except for type of mooring used.
—	Intended to be same as OMA I but with a mechanical disarming device added.
—	Intended to be same as OMA I but with electrical disarming device added.
—	A special assembly for use in depths up to 925 ft., utilizing the OMA I case and EMC anchor and a specially designed float.
1800	Manufacture of this mine was halted prior to 1940, the UMB being considered more suitable. Stocks on hand were laid during World War II. Used preset-type anchor.
1400	Replaced UMA. Uses normal plummet-type anchor.
1400	Differs only in type of mooring used.
1400	Snag line added to defend against shallow-draft vessels.
1400	One cutter mounted high and one deep.

www.ingramcontent.com/pod-product-compliance
Lightning Source LLC
Chambersburg PA
CBHW042032150426
43200CB00002B/30